Wolfgang Woiwode

Qualification of the airborne FTIR spectrometer MIPAS-STR and study on denitrification and chlorine deactivation in Arctic winter 2009/10

Qualification of the airborne FTIR spectrometer MIPAS-STR and study on denitrification and chlorine deactivation in Arctic winter 2009/10

Qualification of the airborne FTIR spectrometer MIPAS-STR and study on denitrification and chlorine deactivation in Arctic winter 2009/10

by
Wolfgang Woiwode

Dissertation, Karlsruher Institut für Technologie (KIT)
Fakultät für Chemie und Biowissenschaften
Tag der mündlichen Prüfung: 19. April 2013
Referenten: Prof. Dr. M. Olzmann, Prof. Dr. J. Orphal

Impressum

Scientific
Publishing

Karlsruher Institut für Technologie (KIT)
KIT Scientific Publishing
Straße am Forum 2
D-76131 Karlsruhe

KIT Scientific Publishing is a registered trademark of Karlsruhe
Institute of Technology. Reprint using the book cover is not allowed.

www.ksp.kit.edu

Print on Demand 2013

ISBN 978-3-7315-0077-3

Qualification of the airborne FTIR spectrometer MIPAS-STR and study on denitrification and chlorine deactivation in Arctic winter 2009/10

Zur Erlangung des akademischen Grades eines
DOKTORS DER NATURWISSENSCHAFTEN
Fakultät für Chemie und Biowissenschaften
Karlsruher Institut für Technologie (KIT) -
Universitätsbereich

genehmigte

DISSERTATION

von

Helmut Walter Wolfgang Woiwode

aus Köln

Dekan:	Prof. Dr. M. Bastmeyer
Referent:	Prof. Dr. M. Olzmann
Korreferent:	Prof. Dr. J. Orphal
Tag der mündlichen Prüfung:	19.4.2013

’Erst die Möglichkeit, einen Traum zu verwirklichen,
macht unser Leben lebenswert.’
Paulo Coelho, *Der Alchimist*

Meinen lieben Eltern

Summary

A detailed understanding of polar stratospheric ozone chemistry and the related microphysical processes is the key for predictions of the evolution of the ozone layer in the future. The grade of Arctic ozone loss in stratospheric winters is linked to the climate change. Recently, individual unusually cold Arctic stratospheric winters accompanied by extended occurrence of polar stratospheric clouds were observed, approaching record ozone loss levels comparable to Antarctic conditions. Important aspects of Arctic polar ozone chemistry are the processes of denitrification and chlorine deactivation. Denitrification is the process of irreversible redistribution of NO_y species (mainly HNO_3) from stratospheric altitudes above $\sim$18 km via condensation, sedimentation and evaporation, delaying the deactivation of ozone-destroying chlorine species. Depending on the extent of denitrification, the efficiency of the deactivation of ozone-destroying chlorine species via the fast deactivation into $ClONO_2$ is limited.

This work describes the qualification of the airborne FTIR spectrometer MIPAS-STR (Michelson Interferometer for Passive Atmospheric Sounding-STRatospheric aircraft) deployed aboard the Russian high-flying aircraft M55 Geophysica. Using the measurements of MIPAS-STR, detailed studies on the processes of denitrification and chlorine deactivation in the Arctic winter lower stratosphere in early 2010 were carried out. This work was embedded in the integrated European project RECONCILE (Reconciliation of Essential Process Parameters for an Enhanced Predictability of Arctic Stratospheric Ozone Loss and its Climate Interactions).

To allow quantitative studies on chemical and physical atmospheric processes, the MIPAS-STR data-processing chain was optimised and modified. The radiometric calibration of the measurements and the quality of the vertical pointing knowledge was improved and completed by a comprehensive characterisation of the current instrument configuration. A central aspect of this work was the elaboration of a retrieval strategy meeting the demands of highly continuum-affected spectra from the UTLS (Upper Troposphere/Lower Stratosphere) region to allow the quantitative determination of atmospheric parameters (i.e. temperature and mixing ratios of O_3, HNO_3, $ClONO_2$, chlorofluorocarbons (CFCs) and H_2O) from the calibrated measurements.

The retrieved vertical profiles from MIPAS-STR are combined to 2-dimensional vertical cross-sections of trace gas mixing ratios and temperature along flight track. For the first time it is proven that from the MIPAS-STR measurements mesoscale atmospheric structures and filaments with vertical extensions in the order of one kilometer and horizontal extensions of several tens of kilometers along flight track can be resolved. Furthermore, it is demonstrated that quantitative retrievals from MIPAS-STR measurements from inside partially transparent polar stratospheric clouds are possible, giving insight into the chemical gas phase composition inside and below these clouds. The retrieval results are validated with collocated independent in-situ measurements.

Vertical cross-sections of trace gas mixing ratios, temperature and cloud information are presented for three RECONCILE flights between the end of January and the begin of March 2010. The measurements show narrow structures and filaments of polar vortex and extra-vortex air, and patterns owing from chemical and physical processing in the vortex air. The retrieval results of HNO_3 and $ClONO_2$ are correlated with retrieved CFC-12, allowing comparisons of similar airmasses from different phases of the Arctic winter. The comparisons show the vertical redistribution of HNO_3 through denitrification and the deactivation of active chlorine species into the reservoir species $ClONO_2$. An important part of the lower stratospheric chlorine budget is analysed by comparing MIPAS-STR measurements of

ClONO$_2$ with in-situ measurements of daytime ClO, with this species being a driving component in catalytic chemical ozone loss. The presented results may be used for the verification of atmospheric chemistry models.

Polar stratospheric clouds are known to play an important role in the polar winter stratosphere by providing sites for the heterogeneous activation of ozone-depleting chlorine species. Certain polar stratospheric cloud particles irreversibly redistribute HNO$_3$ (denitrification) and thereby delay ozone recovery in polar spring. Furthermore, the radiative properties of polar stratospheric cloud particles influence the atmospheric circulation.

Large-dimension potentially HNO$_3$-containing particles inside polar stratospheric clouds potentially capable of denitrification were indicated by in-situ measurements during RECONCILE. These particles can hardly be explained with current theory of denitrification assuming compact spherical particles. In this work, a combined study with simulations from the CLaMS model (Chemical Lagrangian Model of the Stratosphere) and MIPAS-STR measurements at the end of the sensitive cold phase of the Arctic winter in January 2010 was carried out to test possible explanations for the observed particle sizes. Based on the assumptions that these particles were (i) significantly aspheric or (ii) 'flake-like' spherical particles, both characterised by reduced settling velocities, denitrification of large micron-sized NAT (nitric acid trihydrate) particles was simulated with CLaMS. The simulated HNO$_3$ redistribution showing characteristic vertical patterns was compared with the MIPAS-STR measurements. The results of this study indicate that denitrification by large spherical 'flake-like' particles is unlikely in the context of the simulations and observations, while the assumption of compact aspheric particles with moderate aspect ratios (i. e. ratio of height to diameter) is supported. Further in-situ measurements with advanced techniques are required to determine the properties (morphology and composition) of these particles quantitatively. Aspheric NAT particles with moderate aspect ratios might bring the results from in-situ measurements, remote sensing measurements and chemistry transport modelling associated to RECON-

CILE into a context. The consideration of this aspect may help to improve atmospheric chemistry models in the future.

Overall, the MIPAS-STR measurements give insight into specific situations of the Arctic winter stratosphere in early 2010 and provide the basis for further studies on Arctic stratospheric ozone chemistry and related microphysical processes.

Zusammenfassung

Ein umfassendes Verständnis der Chemie des stratosphärischen Ozons und der damit verbundenen mikrophysikalischen Prozesse ist die Voraussetzung für verlässliche Vorraussagen über die Entwicklung der Ozon-Schicht in der Zukunft. Das Ausmaß des Ozonverlustes in arktischen Wintern steht in direktem Zusammenhang mit dem Klimawandel. In den vergangenen Jahrzehnten wurden vereinzelt außergewöhnlich kalte arktische stratosphärische Winter beobachtet, welche von verbreitetem Auftreten von polaren stratosphärischen Wolken und massivem Ozonabbau, vergleichbar mit antarktischen Verhältnissen, geprägt waren. Wichtige Aspekte der polaren Ozon-Chemie sind die Prozesse 'Denitrifizierung' und 'Chlor-Deaktivierung'. Dentrifizierung stellt den Prozess der irreversiblen Umverteilung von NO_y-Spezies (vor allem HNO_3) aus dem Höhenbereich oberhalb $\sim$18 km durch Kondensation, Sedimentation und Evaporation dar, welcher zu einer Verzögerung der Deaktivierung von ozonzerstörenden Verbindungen führen kann. Je nach Ausmaß der Denitrifizierung wird dabei die Effizienz der Deaktivierung von ozonzerstörenden chlorhaltigen Spezies durch die schnelle Überführung in $ClONO_2$ verringert.

Die vorliegende Arbeit beschreibt die Qualifizierung des flugzeug-getragenen FTIR-Spektrometers MIPAS-STR (Michelson Interferometer for Passive Atmospheric Sounding-STRatospheric aircraft), welches auf dem hochfliegenden russischen Forschungsflugzeug M55 Geophysica eingesetzt wird. Weiterhin werden in dieser Arbeit Studien zu den Prozessen Denitrifizierung und Chlor-Deaktivierung unter Verwendung der Messungen von MIPAS-STR in der arktischen unteren Stratosphäre zu Beginn des Jahres 2010 vorgestellt. Die vorliegende Arbeit wurde im Rahmen des integrierten europäischen Projektes RECONCILE (Reconciliation of Essential Process Parameters for an Enhanced Predictability of Arctic Stratospheric Ozone Loss and its Climate Interactions) durchgeführt.

Um die Voraussetzung für quantitative Studien von atmosphärischen Prozessen zu schaffen wurde die Daten-Prozessierung von MIPAS-STR optimiert und modifiziert. Die Qualität der radiometrischen Kalibrierung der Messungen sowie der vertikalen Information der Sichtlinienstabilisierung wurde verbessert. Weiterhin wurde MIPAS-STR in der aktuellen Konfiguration umfangreich charakterisiert. Ein zentraler Aspekt dieser Arbeit war die Ausarbeitung einer Retrieval-Strategie, welche den Anforderungen von stark kontinuum-behafteten Messungen aus dem Höhenbereich der unteren Stratosphäre und oberen Troposphäre gerecht wird. Die ausgearbeitete Retrieval-Strategie erlaubt die quantitative Ableitung von atmosphärischen Parametern (zum Beispiel Temperatur und die Mischungsverhältnisse von O_3, HNO_3, $ClONO_2$, Fluor-Chlor-Kohlenwasserstoffen (FCKWs) und H_2O) aus den kalibrierten Messungen.

Die abgeleiteten Vertikal-Profile werden dabei zu zweidimensionalen vertikalen Querschnitten der Mischungsverhältnisse der jeweiligen Spurengase sowie der Temperatur entlang des Flugweges kombiniert. Dabei wird erstmals gezeigt, dass mit den MIPAS-STR-Messungen mesoskalige Strukturen in der Atmosphäre aufgelöst werden können, zum Beispiel Filamente mit vertikalen Ausdehnungen in der Größenordnung von einem Kilometer und horizontalen Ausdehnungen von einigen zehn Kilometern. Weiterhin wird gezeigt, dass die quantitative Ableitung von atmosphärischen Parametern von Messungen innerhalb teilweise transparenter polarer stratosphärischer Wolken möglich ist. Damit geben die MIPAS-STR-Messungen Einsicht in die chemische Gasphasen-Zusammensetzung innerhalb und direkt unterhalb dieser Wolken. Die Ergebnisse von MIPAS-STR werden dabei mit unabhängigen in-situ-Messungen validiert.

Für drei RECONCILE-Flüge im Zeitrahmen Ende Januar bis Anfang März 2010 werden die abgeleiteten Querschnitte von Spurengas-

Mischungsverhältnissen, Temperatur und Wolken-Bedeckung entlang der Flugwege diskutiert. Die Messungen zeigen schmale Strukturen und Filamente innerhalb und ausserhalb des Polar-Wirbels sowie Muster, welche von der chemischen und physikalischen Prozessierung dieser Luftmassen zeugen. Die Ergebnisse für HNO_3 und $ClONO_2$ werden mit den Ergebnissen für FCKW-12 korreliert, wodurch der Vergleich von ähnlichen Luftmassen in verschiedenen Phasen des arktischen Winters ermöglicht wird. Die Ergebnisse zeigen die vertikale Umverteilung von HNO_3 durch Denitrifizierung und die Deaktivierung von aktiven Chlor-Spezies in $ClONO_2$. Ein wichtiger Teil des unteren stratosphärischen Chlor-Budgets wird durch den Vergleich der MIPAS-STR-Messungen von $ClONO_2$ und in-situ-Messungen von ClO unter sonnenbeschienenen Bedingungen analysiert. ClO stellt dabei eine der wichtigsten Verbindungen im Zusammenhang mit dem stratospärischen chemischen Ozon-Abbau dar. Die Ergebnisse stellen eine mögliche Basis für die Verifikation von atmosphärischen Chemie-Modellen dar.

Es ist bekannt, das polare stratosphärische Wolken eine wichtige Rolle in der polaren Winter-Stratosphäre spielen, indem sie Oberflächen für die heterogene Aktivierung von ozonzerstörenden Verbindungen bieten. Bestimmte Partikel in polaren stratosphärischen Wolken führen zu einer irreversiblen vertikalen HNO_3-Umverteilung (Denitrifizierung), was zu einer Verzögerung der Erholung des stratosphärischen Ozons im polaren Frühjahr führt. Weiterhin wirken sich polare stratosphärische Wolken auf den Strahlungshaushalt der Atmosphäre aus, was zu einer Beeinflussung der atmosphärischen Zirkulation führt.

Während RECONCILE wurden unerwartet große potenziell HNO_3-haltige Partikel innerhalb von polaren stratosphärischen Wolken von in-situ-Instrumenten gemessen. Diese Partikel befinden sich nicht im Einklang mit dem aktuellen Verständnis des Denitrifizierungs-Prozesses, wenn ein kompakter sphärischer Aufbau vorausgesetzt wird. Im Rahmen dieser Arbeit wurde eine kombinierte Studie mit Simulationen des CLaMS-Modells (Chemical Lagrangian Model of the Stratosphere) und Messungen von MIPAS-STR am Ende der kalten Phase des arktischen Winters im Januar 2010 durchgeführt, um mögliche Erklärungen für die beobachteten Partikel zu testen. Basierend auf der Annahme dass diese Partikel entweder (i) asphärisch oder (ii) sphärisch und dabei "flocken-artig" aufgebaut sind, wurde mit CLaMS die Denitrifizierung durch mikrometergroße Partikel aus Salpetersäure-Trihydrat simuliert. Beide Annahmen wurden durch eine charakteristische Reduzierung der Fallgeschwindigkeiten der jeweiligen Partikel modelliert. Die resultierenden simulierten Umverteilungen von HNO_3 zeigen charakteristische vertikale Muster und werden mit den MIPAS-STR-Messungen verglichen. Die Ergebnisse dieser Studie zeigen, dass im Rahmen der Modell-Bedingungen eine Dentrifizierung durch "flocken-artige" Partikel unwahrscheinlich ist, während das Vorhandensein von kompakten asphärischen Partikeln mit moderaten Höhen-/Durchmesser-Verhältnissen unterstützt wird. Weitere in-situ-Messungen mit neuartigen Techniken sind erforderlich, um die Eigenschaften dieser Partikel quantitativ zu bestimmen. Das Vorhandensein von asphärischen Partikeln aus Salpetersäure-Trihydrat mit moderaten Höhen/Durchmesser-Verhältnissen könnte die Ergebnisse von in-situ-Messungen, Fernerkundungs-Messungen und Chemie-Transport-Simulationen im Zusammenhang mit RECONCILE in Einklang bringen. Die Berücksichtigung dieses Aspekts könnte damit Chemie-Modelle der Atmosphäre in Zukunft verbessern.

Insgesamt geben die Messungen von MIPAS-STR Einblick in spezifische Situationen in der arktischen Winter-Stratosphäre im Frühjahr 2010 und bieten damit die Basis für weitere Studien über die Ozon-Chemie der arktischen Stratosphäre und die damit verbundenen mikrophysikalischen Prozesse.

Contents

1 Introduction

1.1 Motivation

Polar stratospheric ozone chemistry remains an important issue of public and environmental interest (Solomon [1999]). In the 1980s, severe catalytic ozone destruction was observed in the Antarctic spring stratosphere, leading to the so-called ozone hole and thereby to the reduction of the overall ozone layer of the earth's atmosphere. In the Antarctic polar vortex, cold temperatures and specific physical and chemical processes were found to support dramatic catalytic ozone depletion. Man-made chlorofluorocarbons (CFCs) were identified to release catalytic chlorine species capable of efficient ozone destruction. In the Arctic, similar processes were observed, however resulting in less extensive and more variable ozone depletion due to different meteorological conditions. In 1989, emissions of chlorofluorocarbons (CFCs) were restricted by the protocol of Montreal, and the increase of ozone destroying species in the stratosphere was stopped. As a consequence, a recovery of the ozone layer was predicted for the next decades.

Today, global climate change going along with decreasing stratospheric temperatures and changes in the atmospheric composition complicate predictions of the evolution of the ozone layer in the future. The level of ozone reduction at Arctic latitudes shows large interannual variability. Frequently, cold Arctic stratospheric winters accompanied by considerable ozone depletion are observed, and a potential trend towards colder temperatures was identified (Rex et al. [2006]). In the recent Arctic winter 2010/11, unusually cold temperatures at stratospheric altitudes resulted in highest levels of column ozone reduction approaching Antarctic conditions (Manney et al. [2011]). These observations point out that a detailed understanding of the manifold and interrelated processes involved in Arctic stratospheric ozone depletion is required to allow meaningful predictions for the evolution of the ozone layer in the future.

While the fundamental ozone-destroying processes in the stratosphere are understood to a high degree, quantitative understanding is lacking for many individual chemical and physical processes (von Hobe et al. [2012]) as well as for aspects of atmospheric dynamics and the linkage to climate change (Sinnhuber et al. [2011]). Current key issues of Arctic stratospheric research are for example:

- the activation and deactivation of ozone destroying catalytic species

- the efficiency of ozone destroying catalytic cycles

- the nucleation process, composition and morphology of certain polar stratospheric cloud (PSC) particles

- the process of denitrification[1]

- mixing of polar vortex air with extra-vortex air and transport through the vortex edge

- the dynamics of the polar vortex, the linkage to the large scale circulation and the influence of planetary wave activity in the context of climate change

The work in hand was carried out in the context of the EU-FP7 project RECONCILE (Reconciliation of Essential Process Parameters for an Enhanced Predictability of Arctic Stratospheric Ozone Loss and its Climate Interactions, see https://www.fp7-reconcile.eu) addressing issues of Arctic

---

[1] Denitrification is the irreversible vertical redistribution of NO_y species through condensation, sedimentation and evaporation of HNO_3-containing particles. The term NO_y encompasses all forms of inorganic nitrogen species in the atmosphere except for N_2 and N_2O.

stratospheric ozone chemistry (von Hobe et al. [2012])). The goal of REC-ONCILE was to enhance the understanding of Arctic stratospheric ozone depletion to allow more reliable predictions for the evolution of the ozone layer in the future. To achieve this goal, RECONCILE involved laboratory studies, chemistry modelling and field campaigns, including flights of the Russian research aircraft M55 Geophyisca capable of probing the Arctic polar stratosphere.

Utilising measurements by the Fourier-Transform Infra-Red (FTIR) spectrometer Michelson Interferometer for Passive Atmospheric Sounding - STRatospheric aircraft (MIPAS-STR) (Piesch et al. [1996]) aboard the high-flying aircraft Geophysica, the aim of this work was to study the processes 'denitrification' and 'chlorine deactivation' in the Arctic winter 2009/10 for specific flights in detail. MIPAS-STR is tailored especially to the sensing of ozone relevant trace gases (i.e. O_3, HNO_3, $ClONO_2$ and CFCs) and allows for the reconstruction of vertically and horizontally resolved distributions of these species along the flight track. Advantages of this measurement technique are that many trace gas species as well atmospheric temperature and information on cloud-coverage are covered simultaneously, and measurements are possible during night and day. Further advantages compared to similar satellite- and balloon-borne techniques are that the sampled airmasses are closer, resulting in lower uncertainties in the retrieval (i.e. through lower requirements for pointing accuracy and local thermodynamic equilibrium conditions at the atmospheric altitudes sampled primarily). Furthermore, defined atmospheric areas can be probed with rather high sampling density and atmospheric regions can be targeted flexibly under consideration of actual meteorological forecasts.

In terms of accuracy and vertical resolution, MIPAS-STR is situated in between comparable satellite techniques and in-situ instruments. In this work it is shown, that mesoscale[2] vertical and horizontal structures along

---

[2]Structures and filaments with vertical extensions of hundreds of meters to a few kilometers and horizontal extensions of a few tens to hundreds of kilometers

flight track can be resolved by the MIPAS-STR configuration with high accuracy. Thereby, atmospheric structures and filaments with vertical extensions in the order of 1 km and horizontal extensions of a few tens of kilometers along flight track are resolved. Such a high vertical and horizontal resolution is especially of interest for comparisons with high resolution models of the atmosphere.

Involving MIPAS-STR measurements from three flights of the Geophysica between the end of January and the begin of March 2010, denitrification and chlorine deactivation are discussed. By analysing complementary MIPAS-STR measurements of $ClONO_2$ and in-situ measurements of ClO, the extent of chlorine deactivation into $ClONO_2$ in the sensitive phase between end of January and begin of March 2010 is quantified. Finally, a combined study involving simulations of the chemical transport model CLaMS (Chemical Lagrangian Model of the Stratosphere, Grooß et al. [2005]) addresses the properties of large potentially HNO_3-containing particles, which were measured in-situ during RECONCILE and are probably triggering denitrification (von Hobe et al. [2012]). The results of this study suggest improvements for chemical transport and climate chemistry modelling.

Important preconditions for these studies were the optimisation of the MIPAS-STR data processing chain and a comprehensive characterisation of the measurements. Furthermore, a retrieval strategy allowing accurate retrievals from aerosol- and cloud-affected spectra from the Upper Troposphere/Lower Stratosphere (UTLS) region probed by MIPAS-STR was elaborated. The retrieval results were comprehensively characterised and validated with collocated independent in-situ measurements.

The introduction of this work proceeds with a short overview of stratospheric ozone chemistry with focus on the Arctic stratosphere, followed by a brief introduction into atmospheric infrared limb-emission spectroscopy. The instrument MIPAS-STR in its current configuration and the carrier M55 Geophysica are introduced in chapter 2. The data-processing chain and the calibration, characterisation and validation of the measurements and the retrieval results are described in chapter 3. Thereby, measurements from the RECONCILE and the ESSenCe campaign (ESa Sounder Campaign 2011, see https://www.essence11.org/Essence) are considered. The retrieval results related to three RECONCILE flights are presented in chapter 4, and the observed mesoscale atmospheric structures and filaments are discussed. Denitrification and chlorine deactivation in the Arctic winter 2009/10 are discussed by bringing the flights into a quantitative context. Finally, the study on denitrification by large-dimension HNO_3-containing particles involving CLaMS simulations and the MIPAS-STR retrieval results is presented. The key results of this work are summarised in chapter 5 together with a perspective for future work.

1.2 The ozone layer and chemistry of the Arctic winter stratosphere

This section discusses the importance of the ozone (O_3) layer in the stratosphere (i.e. the layer of the earth atmosphere between about 15 - 50 km, depending on season and latitude) in general and focusses on ozone chemistry and related microphysical processes in the polar winter stratosphere. Only examples of important references addressing the history of the discovery of the ozone layer, its relevance for the ecological system and the chemical processes leading to ozone depletion can be given in this context. A more detailed overview on atmospheric ozone depletion is provided in the review article of Solomon [1999] (and references therein), which serves as a guideline for this section.

The ozone layer is of essential importance for the existence of life on our planet in its today's form and variety, shielding the earth surface from ultraviolet (UV) radiation capable of destroying cellular tissue. Stratospheric ozone absorbs UV radiation at wavelengths <310 nm (Hartley [1880]) under formation of molecular oxygen (O_2) and atomic oxygen (O) and largely determines the thermal structure of the stratosphere.

First observations of integrated column ozone in the atmosphere began in the early twentieth century (Dobson et al. [1930] and references therein). The vertical structure of the world-wide ozone layer peaking at altitudes of $\sim$30 km was first explained by Chapman [1930]. He showed that molecular oxygen and ozone exchange with each other in an altitude-dependent steady state equilibrium, involving the photolysis of O_3 and O_2 and reactions with atomic oxygen. The so-called 'Chapman cycle' explains the vertical structure of the ozone layer reasonably well in a qualitative way. However, for a quantitative understanding chemical interactions of ozone with other chemical families in the atmosphere must be considered, such as hydrogen oxides, nitrogen oxides and halogen species.

Especially halogen species (chlorine and bromine) destroy ozone efficiently through catalytic cycles. The most important source for ozone-destroying chlorine are man-made chlorofluorocarbons (CFCs) (Molina and Rowland [1974]). The chemical processes resulting from the emissions of CFCs led to the discovery of the Antarctic 'ozone hole' in the 1980s (Farman et al. [1985]) and resulted in a significant thinning of the overall ozone layer. Almost the entire net ozone depletion in the polar winter stratosphere is attributed to catalytic cycles involving chlorine (i.e. Stolarski and Cicerone [1974]; McElroy et al. [1986]).

CFCs are supplied to the stratosphere in the tropics and are then transported to the high latitudes by the Brewer-Dobson Circulation (Brewer [1949]; Holton et al. [1995]). These species are characterised by long atmospheric lifetimes (i.e. $\sim$50 years for CFC-11) due to non-solubility in water (and therefore no wash-out through rain) and chemical inert character. The CFCs are decomposed only slowly by UV photolysis and thereby release chlorine into the stratosphere, which then takes part in catalytic ozone depletion. During most time of the year stratospheric chlorine is trapped in reservoir species (mainly HCl and $ClONO_2$). However, in the Arctic and

Antarctic winter diabatic cooling of the down-welling airmasses inside the polar vortex results in sufficiently cold temperatures for polar stratospheric cloud (PSC) formation (i.e. Peter [1997]). The polar vortex is an extended low-pressure system which forms every winter over the polar regions. Inside the polar vortex characteristic physical and chemical processes take place and affect stratospheric ozone.

Polar stratospheric clouds play an essential role in the polar winter stratosphere: First, the cloud particles provide reactive surfaces for heterogenous reactions, in particular the activation of chlorine reservoir species. Second, certain HNO_3-containing PSC particles irreversibly redistribute HNO_3, which has considerable consequences for the deactivation of ozone-destroying components. Third, the radiative properties of PSC particles (i. e. absorption, emission and scattering of light) affect stratospheric temperatures and thereby influence the atmospheric circulation.

Depending on the ambient temperatures different particle species including (i) ice, (ii) supercooled ternary solutions (STS), consisting of H_2O, HNO_3 and H_2SO_4, and (iii) nitric acid trihydrate (NAT) are formed, while also further species are possible (i.e. Peter and Grooß [2012] and references therein). These species provide reactive surfaces that allow the heterogeneous activation of the chlorine reservoir species according to:

$$ClONO_2(g) + HCl(s) \longrightarrow Cl_2(g) + HNO_3(s) \qquad \text{(R 1)}$$

$$ClONO_2(g) + H_2O(s) \longrightarrow HOCl(g) + HNO_3(s) \qquad \text{(R 2)}$$

$$HOCl(g) + HCl(s) \longrightarrow Cl_2(g) + H_2O(s) \qquad \text{(R 3)}$$

Then, under the twilight conditions of the on-setting Arctic/Antarctic spring, Cl_2 is photolysed and enters the ozone-destroying catalytic cycles. About 75 % of polar stratospheric ozone loss is attributed to the ClO dimer cycle (Molina and Molina [1987]):

$$ClO + ClO + M \longrightarrow ClOOCl + M \qquad (R\,4)$$

$$ClOOCl + h\nu \longrightarrow Cl + ClOO \qquad (R\,5)$$

$$ClOO + M \longrightarrow Cl + O_2 + M \qquad (R\,6)$$

$$2\,(Cl + O_3 \longrightarrow ClO + O_2) \qquad (R\,7)$$

$$\text{Net} \quad 2O_3 + h\nu \longrightarrow 3O_2$$

Chlorine-induced catalytic ozone depletion continues into spring until the active chlorine species are deactivated into their reservoir species. Deactivation into $ClONO_2$ is an important fast reaction channel limiting ozone depletion, while deactivation into HCl occurs more slowly (i.e. McConnell and Jin [2008]):

$$ClO + NO_2 + M \longrightarrow ClONO_2 + M \qquad (R\,8)$$

$$Cl + CH_4 \xrightarrow{\;O_2\;} HCl + CH_3O_2 \qquad (R\,9)$$

$$Cl + HCHO \xrightarrow{\;O_2\;} HCl + HO_2 + CO \qquad (R\,10)$$

The efficiency of the fast deactivation of active chlorine into $ClONO_2$ depends on the availability of NO_2, which in turn is supplied from photolysis of HNO_3. However, the availability of HNO_3 at the altitudes of the ozone layer in polar winters may be limited due to condensation of this species into PSC particles and subsequent sedimentation. This process is named denitrification and is thought to involve PSC particles consisting of HNO_3-containing hydrates (i.e. NAT). Thereby, particle sizes in the order of several microns and characterised by sufficiently high settling velocities are needed to explain vertical patterns of HNO_3 redistribution observed in the polar winter stratosphere. Large HNO_3-containing particles capable of denitrification were observed by Fahey et al. [2001]. However, the composition and morphology of large HNO_3-containing particles involved in denitrification are not yet fully understood (von Hobe et al. [2012]).

In typical Arctic winters, the fast deactivation of chlorine into $ClONO_2$ prevents higher ozone loss rates due to moderate denitrifcation and therefore sufficient availability of NO_2 from HNO_3 photolysis. In contrast, in the Antarctic winter stratosphere cold temperatures provide condensed particles as sites for chlorine activation long into spring-time due to specific meteorological conditions. The associated almost complete denitrification at certain altitudes hinders fast deactivation of active chlorine into $ClONO_2$, allowing catalytic ozone depletion to persist further into spring. The consequence is a dramatic ozone reduction in the order of 50 % of the total ozone column inside the Antarctic polar vortex during each winter/spring season. Thereby, local ozone loss at altitudes affected by PSCs can exceed 90 %. In the Arctic these processes are less severe, yielding variable ozone loss rates ranging from a few percent to a few ten percent, still inducing a significant net reduction of atmospheric ozone when the ozone-poor polar vortex breaks up and the vortex air is mixed with other airmasses. Today it is estimated that total atmospheric ozone has been reduced by 10 % due to man-made emissions. However, locally the ozone loss is much more extensive, such as at the high latitudes in polar spring.

Due to the limitation of the emissions of CFCs by the Montreal Protocol in 1989 the further increase of stratospheric chlorine was stopped. CFCs were mainly replaced by hydrogenated chlorofluorocarbons (HCFCs) and further substitutes, which have much smaller effects on stratospheric ozone. A recovery of the ozone layer to the situation before the 1970s is expected within the next 2-4 decades. However, global climate change and changes in the atmospheric trace gas composition are accompanied by observations of frequent cold Arctic stratospheric winters with considerable ozone destruction (Rex et al. [2006]). A record ozone loss was observed in the recent Arctic winter 2010/11, with local ozone reduction close to Antarctic conditions (Manney et al. [2011]).

The increase of atmospheric CO_2 results in colder stratospheric temperatures, favoring enhanced PSC formation in Arctic winters and thereby leading to enhanced ozone loss (Sinnhuber et al. [2011]). Simultaneously, the increase of CO_2 enhances the Brewer Dobson circulation and planetary wave activity, potentially leading to less stable polar vortices with less severe ozone depletion. Accordingly, the interaction of polar ozone loss with climate change is an important issue of current research (von Hobe et al. [2012]).

Thereby, a detailed understanding of stratospheric ozone depletion chemistry and the related microphysical processes is required to allow meaningful predictions for the development of the ozone layer in the future.

1.3 Atmospheric limb-emission spectroscopy

The principle of FTIR limb-emission spectroscopy is based on the detection of the wave-number dependent distribution of the thermal radiation emitted by molecules in the earth's atmosphere. Measurements are carried out in limb-mode versus the background of the cold space and are converted into spectra. From the spectra, vertical profiles of trace gas volume mixing ratios (VMR), temperature and cloud/aerosol coverage are reconstructed. Upward-viewing measurements provide limited information of these parameters above the flight path. Figure 1.1 illustrates schematically the limb-viewing and upward-viewing measurements performed by MIPAS-STR.

The left panel of Figure 1.2 shows calibrated spectra from a typical limb sequence. The most prominent spectral signatures are labelled accordingly. Further spectral signatures (i.e. from H_2O and $ClONO_2$) are hidden below the dominating signatures. An associated retrieved vertical profile of HNO_3 is shown in the right panel of Figure 1.2. The retrieved vertical distribution of HNO_3 hints on chemical and dynamical structures in the UTLS region. This chapter gives a brief introduction into the theory of

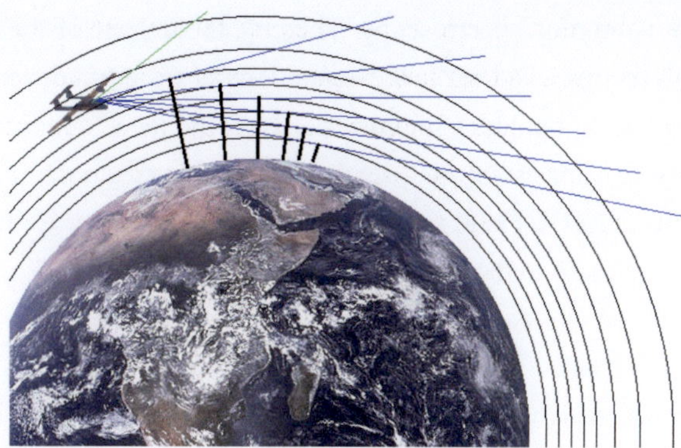

Figure 1.1: Illustration of MIPAS-STR atmospheric measurements in limb-mode (blue) and upward-viewing mode (green). Tangent point positions with respect to the earth surface indicated by black vertical lines (not to scale).

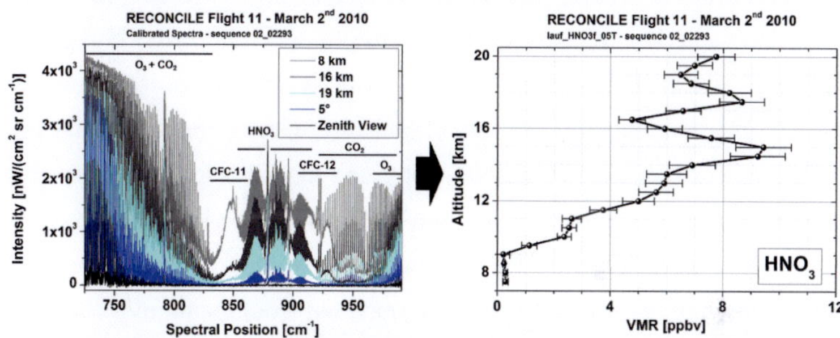

Figure 1.2: Left side: Subset of calibrated MIPAS-STR spectra corresponding to a typical limb scan from a flight altitude of ∼19 km (from Woiwode et al. [2012], with modifications). The approximate regions covered by the most prominent spectral signatures are labeled. Inset: Tangent altitudes of limb observations and elevation angles of upward-viewing observations (0° corresponding to horizontal view). Zenith view corresponding to 90° calibration measurement towards cold space. Right side: Associated retrieved vertical profile of HNO_3 with combined 1σ-error.

thermal limb-emission spectroscopy. Thereby, the linkage of the measured spectra with the retrieved vertical distributions of trace gases, temperature and cloud/aerosol continua is discussed. Here, the theoretical background is summarised from established textbooks and articles such as Rodgers [2000], Hollas [2003], Flaud and Orphal [2011] and references therein and the reader is referred to these authors for further reading.

Spectral transitions

Spectral emission signatures of trace gases arise from molecules in excited states by the relaxation into lower energetic states. Thereby, photons of the energy

$$E = h \cdot \nu \tag{1.1}$$

are emitted, with h being the Planck constant ($h = 6.62608 \cdot 10^{-34}$ J s) and ν the radiation frequency. The resulting spectral emission lines are measured physically and depend on the macroscopic quantities pressure, temperature, mixing ratio and the chemical environment. Essential preconditions for quantitative spectroscopy are the knowledge of the spectral position, source function, the line-intensity and -width as well as the temperature-dependence of the considered transitions. The spectral position and strength of a transition can be obtained from quantum mechanical models. The populations of the involved energy states are described by thermodynamic concepts, while the shapes of spectral lines are determined by quantum mechanic effects as well as physical effects described by the kinetic gas theory. Alternatively, these parameters can be determined directly or indirectly by laboratory measurements. Spectroscopic parameters are tabulated in databases such as the HITRAN (HIgh resolution TRANsmission) database (i.e. Rothman et al. [2009]) and allow for the determination of atmospheric trace gas mixing ratios and temperature from spectroscopic measurements by radiative transfer calculations.

The source function

In case of thermodynamic equilibrium, when the energies of an ensemble of molecules are distributed thermally on the available energy states, the radiation intensity P of a perfect blackbody source with temperature T can be described by the Planck distribution

$$P(\nu, T) = \frac{2h\nu^3}{c^2} \frac{1}{e^{\left(\frac{h\nu}{kT}\right)} - 1} \tag{1.2}$$

with c being the speed of light ($c = 2.99792 \cdot 10^8$ m s^{-1}) and k the Boltzmann constant ($k = 1.38066 \cdot 10^{-23}$ J K^{-1}). The Planck function describes the radiation emitted by a hypothetic perfect emitter at all frequencies in thermal equilibrium. Real emitters significantly differ from this behavior, and blackbody-like emission is only observed for limited spectral regions. Another important aspect is the Kirchhoff law, saying that in case of thermal equilibrium for a fixed frequency and direction emission and absorption are equal. The assumption of local thermal equilibrium also represents an approximation which is applicable for the measurements discussed in this work. In contrast to solid state emitters, molecules in the gas phase are limited to absorption and emission at certain frequencies due to quantum mechanical properties discussed in the following.

Postitions of spectral transitions

The shape of a molecular spectrum is determined by rotational, vibrational and electronic transitions, as long as no excitations leading to dissociation are considered. The associated energetic niveaus in a stationary system can be described by the time-independent Schrödinger equation

$$\hat{H}\psi_{total} = E\psi_{total} \tag{1.3}$$

with $\hat{H}$ being the Hamilton operator, ψ_{total} the total wave function and E_{total}

the total energy of the quantum mechanical system (i.e. the molecule). According to the Born-Oppenheimer approximation ψ_{total} can be separated into the contributions of motion of the cores and the electrons. Furthermore, the contributions of rotation and vibration can also be separated into the vibrational and rotational components:

$$\psi_{total} = \psi_{rot}\,\psi_{vib}\,\psi_{el} \tag{1.4}$$

with ψ_{rot}, ψ_{vib} and ψ_{el} the rotational, vibrational and electronic wave functions. The total energy is the sum of the individual contributions from rotation, vibration and electronic energy:

$$E_{total} = E_{rot} + E_{vib} + E_{el} \tag{1.5}$$

A molecule with N atoms has 3N degrees of freedom, including 3 degrees of freedom for translation and 3 (2 in case of a linear molecule) for rotation. Accordingly, the number of degrees of freedom associated with vibration is (3N-6) for a non-linear molecule and (3N-5) for a linear molecule. The position or energy of a spectral transition is determined by the energy difference between the involved energy states described by the time-independent Schrödinger equation. Transitions between rotational states have lower energies than vibrational transitions, which in turn have lower energies than electronic transitions. While single rotational transitions are possible, vibrational transitions are always accompanied by rotational transitions, and electronic transitions are accompanied by the latter two types of transition.

In the wave number domain, pure rotational transitions are typically situated in the microwave and far IR region between 1 to 500 cm^{-1} ($\lambda \approx 1$ cm to 20 μm). Vibrational transitions are typically found in the middle and near IR region between 500 to 12000 cm^{-1} ($\lambda \approx 20$ μm to 800 nm) and electronic transitions are typically situated in the ultraviolet and visible region between 12000 to 250000 cm^{-1} ($\lambda \approx 800$ to 40 nm). In this work, pure rotational and rotational-vibrational transitions are important and are introduced in the following.

Rotational transitions

Solving the Schrödinger equation for a hypothetic rigid rotating 2-atomic molecule yields the spherical harmonical wave functions ψ_{rot} with the associated undisturbed energy eigenvalues

$$E_{rot} = J(J+1)\frac{\hbar}{2I} \tag{1.6}$$

with J the rotational quantum number ($J = 0, 1, 2, \ldots$), $\hbar$ the quotient of h and $2{\cdot}\pi$ and I the moment of inertia. In the absence of an external electric or magnetic field the rotational energy levels of a two atomic molecule are $(2J+1)$-fold degenerated. The energy difference between two adjacent rotational niveaus is calculated according to

$$\Delta E_{rot} = 2J\frac{\hbar}{2I} = 2JhcB \tag{1.7}$$

with B being the rotational constant. According to quantum mechanical selection rules, for an infrared-active rotational transition a molecule needs to have a permanent dipole moment, and only transitions with $\Delta J = \pm 1$ are allowed. In contrast to linear molecules, non-linear molecules with more atoms are characterised by two or three main moments of inertia depending on the symmetry. Accordingly, solving the Schrödinger equation yields more energy states here.

Vibrational transitions

Solving the Schrödinger equation for a 2-atomic molecule yields the vibrational wave functions ψ_{vib} and the associated energy eigenvalues

$$E_{vib} = h\nu_0 \left(W + \frac{1}{2} \right) \tag{1.8}$$

depending on the eigenfrequency v_0 and the vibrational quantum number W ($W = 0, 1, 2, \ldots$). An important aspect is that for $W=0$ the vibration energy is not equal to zero. This is explained by the Heisenberg uncertainty principle, saying that it is impossible to determine the location and the momentum of an atom or a molecule with absolute accuracy at the same time. Within the approximation of the harmonic oscillator, the energy eigenvalues are distributed equidistantly and the difference between two niveaus is

$$\Delta E_{vib} = h v_0 = \hbar \sqrt{\frac{D}{\mu}} \tag{1.9}$$

with D being the force constant of the oscillating system and the reduced mass

$$\mu = \frac{m_1 \cdot m_2}{m_1 + m_2} \tag{1.10}$$

with m_1 and m_2 being the masses of the individual atoms. According to quantum mechanical selection rules, for an infrared-active vibrational transition a molecule needs to have a dipole moment varying with time, and only transitions with $\Delta W = \pm 1$ are allowed.

However, the approximation of the harmonic potential does not hold for real molecules, especially in higher excited vibrational states. The vibrational energy levels are not equidistant but are getting closer and closer for higher energy levels, and cleavage of the chemical bond is possible. Furthermore, transitions with $\Delta W = \pm 2, \pm 3, \ldots$ are possible due to anharmonicity and result in overtones, with the corresponding intensities being decreased by factor $\sim$10-100 each.

Line strength

The strength of a molecular transition is determined on the one hand by the occupation probabilities of the two energy states involved and on the other hand by the transition probability. In case of thermal equilibrium, the temperature-dependent line-strength is described by

$$j(T) = \frac{8\pi^3}{4\pi\varepsilon_0 3hc} \, v_{ij} d_i^{-1} e^{-\frac{E_j}{kT}} \left(1 - e^{-\frac{h v_{ij}}{kT}} \right) |R_{ij}|^2 \, Q(T)^{-1} \quad (1.11)$$

with ε_0 the electromagnetic permeability of the vacuum ($\varepsilon_0 = 8.854188 \cdot 10^{-12} C^2 N^{-1} m^{-2}$), v_{ij} the transition frequency, d_i the factor of degeneracy, R_{ij} the transition array element and $Q(T)$ the partition function (Ebert and Wolfrum [2001]).

Line profiles

Since real molecules in the gas phase have different velocities and interact with their surrounding, spectral transitions between two energy states are not observed at a single monochromatic frequency. Instead, a frequency distribution of the radiation intensity is observed around the centre frequency v_{center} of a transition, which is described by a line shape function $g(v - v_{centre})$ normalised to unity:

$$\int_{-\infty}^{+\infty} g(v - v_{centre}) dv = 1 \quad (1.12)$$

Contributions to the observed frequency distribution of a spectral transition basically arise from (i) the natural broadening, (ii) Doppler broadening and (iii) pressure broadening. The natural line width is a consequence of the finite life-time of an excited state and the resulting line shape is described by a Lorentz function. The Doppler broadening of spectral lines is a consequence of the thermal velocity distribution of molecules in the gas phase relative to the observer. Measured radiation from a transition is therefore affected by the Doppler effect. The velocity distribution of the molecules can be described by the Maxwell velocity distribution, which is directly mirrored into the observed line shape in form of a Gaussian profile. Pressure broadening is a consequence of the fact that the life-times of excited states are limited by collisions with other molecules. Collisions with different molecules result in different pressure broadening of the spectral transitions of a specific gas. The resulting net line-shapes contain contributions from all three discussed broadening mechanisms, while the relative importance of the individual broadening effects depends on the conditions of the measurement. The measured line-shapes under atmospheric conditions considered in this work are dominated by contributions from Doppler- and pressure broadening, and the resulting line profiles are described by Voigt profiles. The Voigt profile $g_V(v)$ is the convolution of the Gauss profile $g_G(v)$ arising from Doppler broadening and the Lorentz profile $g_L(v)$ induced by pressure broadening:

$$g_V(v) = \int g_L(v) \otimes g_G(v - v')dv' \tag{1.13}$$

For the Voigt profile no analytic solution is available and therefore approximations are applied.

Radiative transfer equation

Limb observations versus cold space (compare Figure 1.1) integrate radiation arising from atmospheric molecules along the instruments line-of-sight (LOS). The intensities of the observed spectral lines are affected by emission, absorption and also scattering processes along the LOS. Within single limb observations radiation contributions from many different atmospheric layers are detected, giving together rise to the observed spectrum. Since pressure, temperature and chemical composition as well as particle loading are variable along the LOS, the corresponding contributions of radiation along the LOS have to be described adequately. The Chandrasekhar form of the radiation transfer equation describes the spectral radiance $L(v, l_{obs})$ seen at an observer position l_{obs} (i.e. from aircraft) when pointing into the atmosphere

$$L(\tilde{v}, l_{obs}) = L(\tilde{v}, l_0) \tau(\tilde{v}, l_{obs}, l_0) + \int_{l_{obs}}^{l_0} J(\tilde{v}, l) \sigma_{a,total}^{Vol}(\tilde{v}, l) \tau(\tilde{v}, l_{obs}, l) dl \quad (1.14)$$

with $\tilde{v}$ being the wave number[3], $L(\tilde{v}, l_0)$ the radiance of a potential background source at the begin of the optical path l_0, $\tau(\tilde{v}, l_1, l_2)$ the spectral transmission between the two optical path coordinates l_1 and l_2 and $\sigma_{a,total}^{Vol}$ the total volume absorption coefficient. As the limb-emission measurements discussed in this work were measured against cold space the background contribution can be neglected (i.e. the left term on the right side of the equation). In case of thermal equilibrium the source function $J(\tilde{v}, l)$ is equal to the Planck function of the corresponding temperature. The spectral atmospheric transmission is calculated according to

$$\tau(\tilde{v}, l_1, l_2) = e^{-\int_{l1}^{l2} \sigma_{a,total}^{Vol}(\tilde{v}, l) dl} \quad (1.15)$$

The total absorption coefficient per volume $\sigma_{a,total}^{Vol}(\tilde{v}, l)$ is the sum of the absorption coefficients $\sigma_{a,gas}^{Vol}(\tilde{v}, l)$ of the individual trace gases m and the aerosol extinction $\sigma_{e,aero}^{Vol}(\tilde{v}, l)$ with

$$\sigma_{a,gas}^{Vol}(\tilde{v}, l) = \sum_{m=1}^{M} \sigma_{a,m}(\tilde{v}, l) \, \eta_m(l) \quad (1.16)$$

---

[3] Wave number $\tilde{v}$, frequency v and wavelength λ are converted into each other via the relation $\tilde{v} = \dfrac{v}{c} = \lambda^{-1}$. The wavenumber $\tilde{v}$ is often used in IR spectroscopy for representing the energies of spectral transitions.

and

$$\sigma_{e,aero}^{Vol}(\tilde{v},l) = \sigma_{a,aero}^{Vol}(\tilde{v},l) + \sigma_s(\tilde{v},l) \tag{1.17}$$

Here, M is the total number of gases considered, $\sigma_{a,m}(\tilde{v},l)$ the absorption coefficient of the individual gas and η_m the absorber number density of species m. This equation represents the link between the absorption and the abundances of trace gases in the atmosphere. $\sigma_{a,aero}^{Vol}(\tilde{v},l)$ is the aerosol absorption coefficient per volume and $\sigma_s(\tilde{v},l)$ the scattering coefficient.

The absorption coefficient $\sigma_{a,gas}^{Vol}(\tilde{v},l)$ of an individual gas m contains the sum of the absorption cross-sections $\sigma_{a,m}(\tilde{v},l)$ of all relevant transitions n. The absorption cross sections are each the product of the corresponding line strength and line shape function. Especially for more complex molecules these parameters are often not known for the large number of individual transitions and therefore net absorption cross-sections measured in the laboratory and available continuum information are applied.

Retrieval

The retrieval is the process of deriving target parameters from remote sensing measurements. Target quantities can be (i) atmospheric parameters (i.e. temperature or trace gas volume mixing ratios) or instrumental parameters (i.e. line-of-sight offset). The retrieval is started with assuming an initial guess atmosphere (i.e. best estimation of the atmospheric state) and a forward calculation of the associated spectra for the measurement geometries of a limb scan. Then, in the inversion process, the calculated spectra $f(x)$ are fitted iteratively to the measured spectra y (each summarised in a vector) under variation of the target parameters x (also summarised in a vector). In case of convergence and agreement between the measured and calculated spectra the associated retrieved target parameters give a best representation of the atmospheric state or instrumental parameters. Trace gases are retrieved in selected spectral microwindows with high sensitivity for the respective species and low spectral interference with other gases. For forward modelling of the measured spectra, the atmosphere is separated into discrete vertical layers, and the radiation transfer is calculated along discrete path elements. Dedicated radiative transfer models allow the forward calculation of the spectra $f(x)$ associated with the initial guess atmosphere. Also provided are the spectral derivative matrices $\mathbf{K}$, which represent the link between the modelled spectra and the corresponding atmospheric or instrumental target parameters. Thereby, the matrix elements $K_{ij} = \delta F_i(x)/\delta x_j$ are the spectral derivatives for the spectral gridpoint i and the retrieval parameter gridpoint j (i.e. trace gas mixing ratio at the altitude j).

In many cases the inversion problem is underestimated due to limited information contained in the measurements and spectral noise. Therefore, side constraints are applied, introducing a priori knowledge into the inversion process and allowing stable solutions. A suitable side constraint should introduce as little a priori knowledge as possible, allowing the extraction of the largest possible fraction of information from the measurement. The specific forward modelling and inversion procedure applied for the MIPAS-STR measurements is discussed in Section 3.2.2 in more detail.

2 Overview MIPAS-STR

The airborne FTIR spectrometer MIPAS-STR is designed for the detection of limb emission spectra of atmospheric trace gases in the mid-infrared. The measurements of MIPAS-STR allow the reconstruction of vertical profiles and cross-sections of atmospheric trace gases, temperature and cloud/aerosol coverage along the flight path. This section introduces the airborne platform M55 Geophysica, the instrument MIPAS-STR and the sampling characteristics.

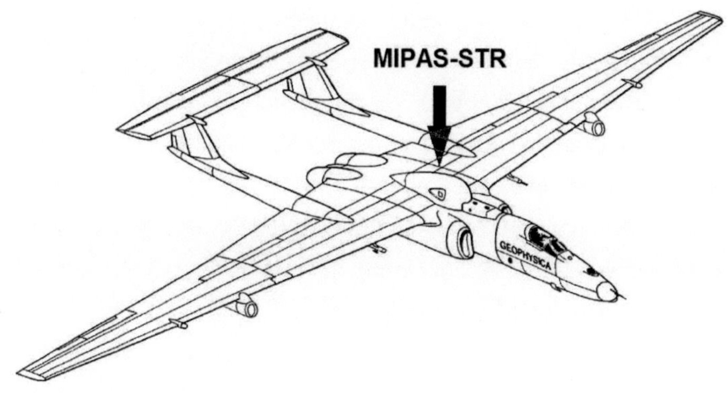

Figure 2.1: MIPAS-STR (hidden below cowling) aboard the Russian high altitude research aircraft M55 Geophysica (from MDB [2002], with modifications).

Table 2.1: Technical data of the M55 Geophysica (MDB [2002]).

Wingspan	37,5 m
Length	22.9 m
Height	4.83 m
Max. payload	~2000 kg
Max. takeoff weight	~24.5 tons
Max. airspeed	750 km/h
Max. flight altitude	20 km
Operating range	~3000 km
Crew	1
Engines	2 turbofan engines PS-30V12 (2x50 kN)

MIPAS-STR is deployed aboard the Russian high altitude research aircraft M55 Geophysica, capable of flight altitudes of 20 km and flight ranges of about 3000 km. The instrument is mounted on top of the aircraft, pointing perpendicular to the right hand side of the flight path (Figure 2.1).

The aircraft type M55 (NATO reporting name: Mystic-B) was designed by Myasishchev Design Bureau in the Soviet Union in the late 1980s as a high altitude reconnaissance aircraft (MDB [2002]). In the 1990s, the M55 set up several world records in flight altitude. Since the late 1990s, after the breakdown of the Soviet Union, the modified research version M55 Geophyisca is deployed for scientific missions. The capability of the M55 Geophysica of carrying payloads of ~2000 kg up to stratospheric altitudes makes it an ideal carrier for MIPAS-STR. Characteristics of the M55 Geophysica are summarised in Table 2.1.

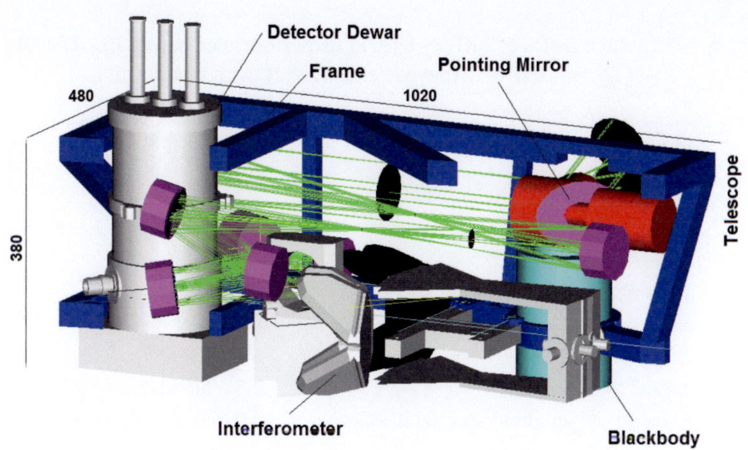

Figure 2.2: Schematic representation of the MIPAS-STR optics module (from Blom et al. [1998]). Mirrors are shown in pink. The light path from the atmosphere via pointing mirror, telescope and interferometer into the detector dewar is indicated by green lines. Dimensions in mm.

2.1 Instrument overview

Technical characteristics of MIPAS-STR are described in detail in Piesch et al. [1996], Kimmig [2001] and Keim [2002]. An overview on the setup, performance and data processing of MIPAS-STR during the RECONCILE and ESSenCe campaigns is given in Woiwode et al. [2012] and is summarised in the following. MIPAS-STR basically consists of an optics and an electronics module. The optics module includes the scan mirror, a mirror telescope, the interferometer and the detector dewar unit. To minimise instrumental background radiation, the optics module is cooled by dry-ice to about 210 K and mirrors are coated with gold or Silflex MK2^{TM}, respectively. The MIPAS-STR optics module is represented schematically in Figure 2.2 and the characteristics of MIPAS-STR in its performance during the RECONCILE and ESSenCe campaigns are summarised in Table 2.2.

Table 2.2: Characteristics of MIPAS-STR in its performance during RECONCILE and ESSenCe (from Woiwode et al. [2012], with modifications).

Telescope	
Field of view (full cone)	$0.44°$
Etendue	2.6×10^{-3} sr cm^2
Interferometer	
Effective optical path difference	± 13.9 cm
Scan time per interferogram	$\sim$9.5 sec
Sampling frequency	48.8 kHz
Signal frequency	2.3 – 5.8 kHz
Unapodised/apodised spectral resolution	0.036 cm^{-1}/0.058 cm^{-1}
Detectors	
NESR (single apodised spectrum, in-flight)	
Channel 1 (725 – 990 cm^{-1})	10 nW/(cm^2 sr cm^{-1})
Channel 2 (1150 – 1360 cm^{-1})	8 nW/(cm^2 sr cm^{-1})
Channel 3 (1560 – 1710 cm^{-1})	5 nW/(cm^2 sr cm^{-1})
Channel 4 (1810 – 2100 cm^{-1})	5 nW/(cm^2 sr cm^{-1})
Pointing	
Pitch/roll accuracy (AHRS)	0.5 arcmin (1σ)
Yaw accuracy (AHRS)	$0.3°$ (1σ)
Estimated LOS-elevation accuracy	1.25/0.78 arcmin (1σ)
Dimensions	
Optics module	135 x 75 x75 cm^3
Electronics module	50 x 50 x 50 cm^3
Total mass	$\sim$200 kg
Power consumption	$\sim$300 W (28 V DC)

Thermal atmospheric radiation from defined elevation angles enters the instrument via the actively controlled and stabilised scan mirror and is then directed into the telescope. The 3-mirror telescope suppresses radiation from outside the field-of-view (FOV), which is scattered at surfaces inside the instrument or diffracted at the edges of the front optics. The core of the instrument is the double pendulum interferometer (Fischer and Oelhaf [1996]), a variation of the classical Michelson interferometer. The interferometer is basically set up by two moving cube corner mirrors connected by the pendulum structure, two flat mirrors, a kalium bromide beamsplitter and the focussing optics towards the detector. Incoming collimated radiation is split up into two parts by the beamsplitter, with each part passing different pathways of variable length. After recombination, the radiation is focussed at the IR detector. By sweeping the pendulum in the horizontal plane around the pendulum axis, the optical pathways of the two light fractions are varied, leading to variable interference for different wavelengths of radiation. During a complete interferometer sweep, an interferogram consisting of the interference pattern arising from all wavelengths transmitted to the detector in dependence of the mechanical pathway is detected. By inverse Fourier Transformation the recorded interferograms in the path domain are transformed into spectra in the frequency domain, which are needed for further processing. A HeNe-Laser ($\lambda = 633$ nm) is also coupled into the interferometer for accurate optical path detection and interferogram sampling.

An advantage of the double pendulum interferometer is a rather compact design, allowing however for a considerably high spectral resolution. Thereby, a one-fold mechanical path difference results in an 8-fold optical path difference. The effective optical path difference of the MIPAS-STR interferometer is ±13.9 cm. Double-sided interferograms are recorded and spectra with an unapodised spectral resolution of 0.036 cm^{-1} are yielded. For the Fourier Transformation of the MIPAS-STR spectra the Norton-Beer strong apodisation (Norton and Beer [1976], Norton and Beer [1977]) is applied, resulting in an effective spectral resolution of 0.058 cm^{-1}.

Vibrations are a typical characteristic of an airborne platform such as the M55 Geophysica and result in velocity variations of the interferometer speed. As a consequence so-called 'ghost-lines' arise in the spectra.

The effects of velocity variations on the MIPAS-STR interferometer were investigated by Kimmig [2001]. Thereby, the beamsplitter assembly was improved and a time-equidistant sampling procedure proposed by Brault [1996] was adapted, making the MIPAS-STR interferometer insensitive to vibrations.

After passing the interferometer, the infrared radiation is directed via an anti-reflectance coated ZnSe window into the detector dewar unit. The detector dewar is cooled to $\sim$4K with liquid helium and houses the four Si:As back-illuminated band-impurity detectors of the four channels. Incoming radiation is split up into four parts by dicroitics, and each fraction of radiation is directed through the individual optical bandpass-filter of the corresponding channel (Table 2.2). The splitting into separate channels reduces the photon noise by reducing the number of incoming photons at the individual detectors.

The electronics module of MIPAS-STR is set up by a hierarchic transputer network. Top system is a PC-based computer system running a real-time UNIX® system. The interferometer electronics, the line-of-sight (LOS) electronics and the housekeeping/auxiliary electronics are subsystems, each able to continue servicing in case the top system interrupts working (i.e. in case of reboots of the top system during flight). An automatic sequence control ensures a fully automatic system starting and operation during flight. The system however can be accessed during flight via an IRIDIUM satellite link (see http://www.iridium.com/default.aspx) for data quality control and potential modifications of the atmospheric sampling programme.

Since MIPAS-STR performs limb-emission measurements at fixed tangent altitudes (i.e. fixed elevation angles versus horizon), an accurate pointing stabilisation is required (Keim [2002]). The Attitude and Heading Reference System (AHRS) included in MIPAS-STR is a Schuler-adapted GPS-aided strapdown inertial navigation system and provides precise attitude angles at a data rate of 128 Hz with low latency. The angle of the scan mirror is measured by a 19-bit encoder and is actively stabilized relative to the attitude information of the AHRS, allowing a near real-time pointing stabilisation. The pointing information is corrected post-flight for drifts in the AHRS data, ensuring accurate pointing knowledge for further data processing steps. Due to the exposure of the instrument to large temperature differences between ground and stratospheric altitudes in the order of 50 K (i. e. for Arctic winter flights), slight warping of the optical components versus the coordinate system of the AHRS is observed. The observed warping is typically in the order of a few arcmins and is reproducibly quantified by line-of-sight retrievals for individual flights. The estimated accuracy of the a-posteriori LOS elevation angles is 1.25/0.78 arcmin[1] and considers the accuracy of the AHRS data, the accuracy of the scan-mirror control and uncertainties inherent to the LOS retrieval. The overall LOS error corresponds to about 5 % of the instrumental FOV and an uncertainty of about 200 m at the lowest tangent altitude of 5 km.

2.2 Atmospheric sampling

The MIPAS-STR sampling programme includes atmospheric measurements in limb-mode and upward-viewing mode (Figure 1.1), which are complemented by calibration measurements with a cold blackbody and versus cold space (zenith view (ZV) measurements). A standard sampling programme of MIPAS-STR covering tangent altitudes between 5 km and flight altitude and complemented by comprehensive upward sampling is indicated schematically in Figure 2.3. Two full limb scans are shown, as the scans are arranged in a mirrored pattern. The limb scanning measurements allow for the retrieval of vertically resolved profiles of atmospheric parameters. The upward viewing measurements contain mainly column information on atmospheric constituents above the flight path and also allow resolving ver-

---

[1]The value of 1.25 arcmins represents a more conservative estimation considering for dynamic variations of the LOS offset during flight and potential systematic errors in the retrieval of the systematic LOS correction.

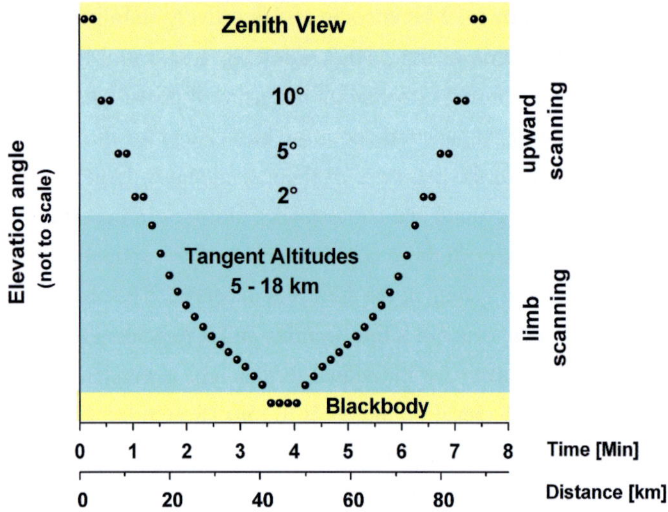

Figure 2.3: Standard sampling programme of MIPAS-STR (from Woiwode et al. [2012]).

tical information to a limited extent. For the shown programme the vertical spacing of the tangent altitudes is 1 km above 8 km altitude and 1.5 km below. An additional limb view with an elevation angle of -0.3° is performed and is characterised by a tangent altitude close to the flight path. Overall, an oversampling of 2-3 is obtained for the limb-views with respect to the FOV opening-angle of 0.44°. The vertical diameter of the FOV increases from about 0.3 km at the highest tangent point to about 3 km at the lowest tangent point.

An important aspect is that the intensity of incoming atmospheric radiation is not homogeneous over the vertical FOV opening-angle, but depends on the vertical distribution of pressure, temperature and trace gases. Due to the exponential increase of pressure towards lower altitudes, the largest radiation contribution by trend arises from the lower part of the FOV. Another important aspect is that identical radiation contributions entering the MIPAS-STR FOV under different vertical angles are not transferred at equal intensity to the detector due to the characteristics of the focussing optics towards the detector. Therefore, a vertical FOV weighting function is determined from characterisation measurements on ground and is considered in retrievals. One full limb scan of the standard sampling programme from a flight altitude of 18 km takes about 3.8 minutes, corresponding to a flight path $\sim$45 km. For the highest limb view, the horizontal distance of the tangent point location from the aircraft position is about 30 km (tangent altitude 17.9 km), increasing to a distance of about 400 km for the lowest limb view (tangent altitude 5 km).

The sampling programme of MIPAS-STR can be flexibly adapted to scientific demands. For example, a modified sampling programme omitting tangent altitudes below 9 km and with less frequent upward viewing measurements can be performed in case higher horizontal sampling density is required or if the lower limb views are affected by optically dense clouds. The duration of a limb scan of the modified sampling programme is about 2.4 minutes, resulting in a horizontal sampling of approximately 25 km.

3 Data processing, characterisation and validation

This chapter describes the processing of the measured raw data towards characterised data products. The level-1 processing covers all calibration steps performed to convert the measured raw interferograms into radiometrically and spectrometrically calibrated spectra with known geolocation. The level-2 processing comprises the reconstruction of atmospheric parameters (temperature, trace gas volume mixing ratios and aerosol distributions) from the calibrated spectra.

In this chapter, the individual data processing steps are introduced briefly and the characterisation results are presented. Modifications in data processing applied within this work and improvements achieved are discussed in more detail. The focus is set on data characterisation, the establishment of a retrieval strategy meeting the demands of highly cloud- and continuum-affected spectra from the UTLS region and the validation of the retrieval products. Key results from the individual steps are summarised at the end of the respective subsections.

3.1 Transformation of raw interferograms into calibrated spectra with known geolocation (level-1 processing)

A reliable radiometric calibration of the MIPAS-STR spectra and accurate pointing knowledge as well as a comprehensive characterisation of the the instrument are essential for quantitative retrievals suited for scientific evaluation. This section describes the MIPAS-STR level-1 processing carried out in this work, the improvements which were achieved, and the char-

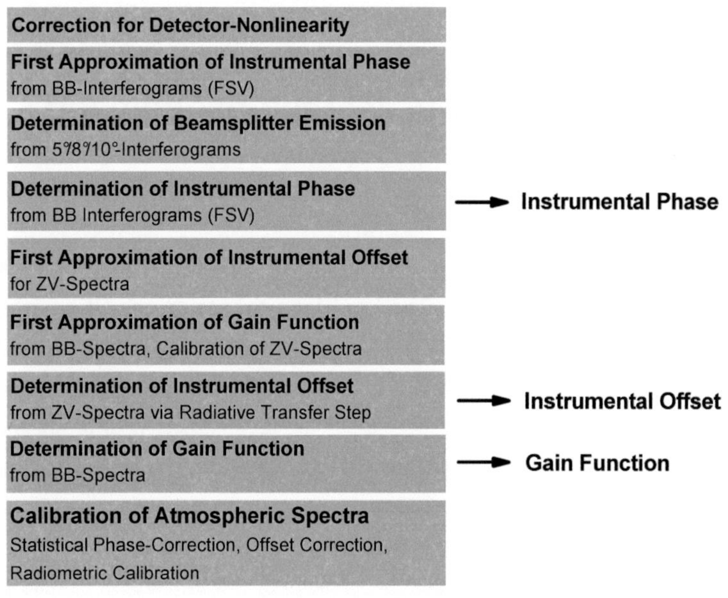

Figure 3.1: Schematic representation of the MIPAS-STR radiometric and spectral calibration procedure (from Woiwode et al. [2012]). BB=blackbody; ZV=zenith view; FSV=phase determination according to Forman, Steel and Vanasse (Forman et al. [1966]).

acterisation of MIPAS-STR its current configuration. The concept of the radiometric and spectral calibration of MIPAS-STR measurements is given by Höpfner et al. [2001b], Kimmig [2001], Keim [2002] and summarised in Woiwode et al. [2012]. The latter reference serves as guideline for this section.

A schematic overview of the calibration steps carried out within the level-1 processing of the MIPAS-STR spectra is given in Figure 3.1. Characteristics to be determined are (i) the nonlinearity in detector response, (ii) the instrumental phase needed for Fourier Transformation of the atmospheric interferograms into spectra, (iii) the radiometric offset of the spectra arising from instrumental background radiation and (iv) the radiometric gain function to convert the atmospheric raw spectra into radiance units. The steps phase correction, determination of instrumental offset and radiometric calibration are carried out separately for the different interferometer sweep directions (forward/backward) as data acquisition is slightly different. For these steps, each flight is separated into sections where the instrumental phase does not change significantly and the instrumental offset and the radiometric gain vary approximately linearly with time. These characteristics are required, since the instrumental phase is averaged and the other two parameters are fitted linearly in time for the atmospheric measurements for calibration. The linear fitting of the instrumental offset and radiometric gain provides accurate assignment of these quantities to the atmospheric measurements and reduces the measurement noise compared to considering only individual calibration measurements. The characterisation of the vertical FOV weighting function and the post-processing and characterisation of the pointing information are independent steps and are discussed separately.

3.1.1 Correction of detector nonlinearity

As the Si:As back-illuminated band impurity detectors show a nonlinear response at increasing photon fluxes, the measured raw interferograms are distorted and the associated spectra show characteristic artefacts. Therefore, the first step is the quantification of the detector nonlinearity. A scaling correction is applied to the interferograms to consider for changes in the mean detector response between spectra with low photon fluxes (i.e. zenith view measurements) and spectra with high photon fluxes (i.e. blackbody measurements) to keep the two-point calibration (Section 3.1.3) valid (distortions *within* the dynamical ranges of the single interferograms are neglible and are not corrected). The approach for detector nonlinearity correction of MIPAS measurements is discussed by Kleinert [2006] and is introduced briefly in the following.

A measured nonlinear raw interogram IFG can be approximated by a polynomial function of linear interferograms:

$$IFG \approx IFG_{lin} + a_2 \, IFG_{lin}^2 + a_3 \, IFG_{lin}^3 \tag{3.1}$$

Consequently, the corresponding uncorrected spectra A_{meas} can be represented by convolutions of the undisturbed spectra A_{lin} with themselves and show higher order artefacts:

$$A_{meas} \approx A_{lin} + a_2 \, A_{lin} \otimes A_{lin} + a_3 \, A_{lin} \otimes A_{lin} \otimes A_{lin} \tag{3.2}$$

Thereby, the observed artefacts can be used to characterise the detector nonlinearity. The iterative minimisation approach for the measured artefacts described by Kleinert [2006] allows for the determination of the parameters a_2 and a_3 and thereby for the correction of the interferograms. Dedicated blackbody calibration measurements without digital filtering for data reduction are utilised for the determination of the detector nonlinearity. Typically between 10 and 30 interferograms of sufficient quality (i.e. low noise and low further electromagnetic disturbances) are obtained within one calibration measurement phase for each interferometer sweep direction.

The measured interferograms from each sweep direction are averaged and converted into spectra which are averaged again. Finally, the observed artefacts in the resulting averaged spectrum are minimised iteratively. In the upper panel of Figure 3.2 a typical channel 1 blackbody spectrum used for detector nonlinearity quantification is shown together with the same spectrum after correction. The quadratic and cubic artefacts are clearly visible, whereas no higher order artefacts are identified. The 30 to 280 cm^{-1} interval is used for the minimisation of the quadratic artefacts and the 2150 to 2900 cm^{-1} interval for minimisation of the cubic artefacts. Remaining signatures in the corrected spectrum outside the spectral range of channel 1

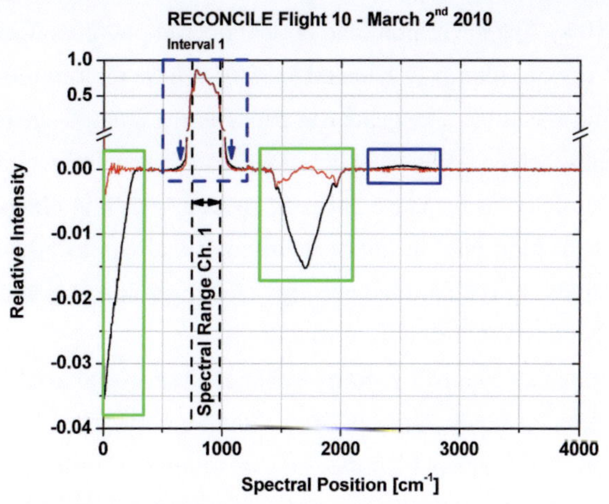

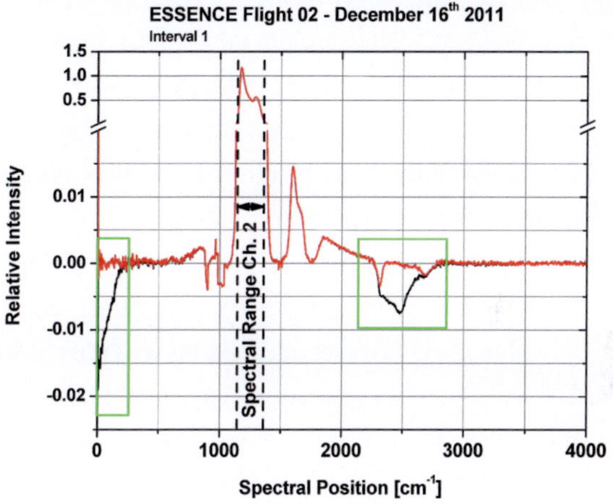

Figure 3.2: Characterisation of MIPAS-STR detector nonlinearity. Uncorrected spectrum (black) and spectrum after minimisation of artefacts (red) for channels 1 (upper panel) and 2 (lower panel). Regions affected by quadratic artefacts marked in green and by cubic artefacts in blue.

in the regions of the quadratic artefacts result from the AC coupling in the signal chain (0 to 30 cm^{-1} region) and optical double-passing effects (1400 to 2100 cm^{-1} region) (compare Kleinert [2006]). These regions are avoided for artefact minimisation. The quadratic artefacts are situated outside channel 1 and a cubic artefact only weakly influences this spectral region. However, the mean detector response between spectra with low photon fluxes and spectra with high photon fluxes is different and has to be corrected. Detector nonlinearity is quantified once for each flight and is assumed to be approximately constant during the entire flight.

The lower panel of Figure 3.2 shows a channel 2 spectrum used for determination of detector nonlinearity prior to and after correction. For channel 2 only quadratic artefacts are identified and are minimised utilising the 20 to 220 cm^{-1} interval. Also here further artefacts attributed to the AC coupling and optical double-passing are observed and affect the quadratic artefact regions. Further signatures are visible around the channel 2 spectral window towards higher and lower wave numbers. These signatures are attributed to optical leakages of the optical band-pass filter (i.e. signature between 1500 and 1750 cm^{-1}) and furthermore result from the electromagnetic disturbances during the calibration measurements from this flight.

The derived nonlinearity parameters for the individual flights are listed in Table 3.1. For channel 1, the nonlinearity parameters are referenced to a common DC value of 2.086 V corresponding to a typical blackbody spectrum. Furthermore, the DC value of a typical zenith view spectrum of 1.480 V is considered to quantify the changes in the detector response between spectra with high and low photon fluxes. Nonlinearity parameters for channel 2 are reported exemplarily for the ESSenCe flight on December 16th 2011[1] and are referenced to a DC value of 1.700 V corresponding to typical blackbody spectrum.

---

[1]For the retrievals discussed in the following only channel 1 spectra were considered. The first complete characterisation of channel 2 in the context of this work is discussed to provide a reference for future work.

Table 3.1: Calibration measurements for detector nonlinearity quantification and derived correction parameters for RECONCILE and ESSenCe flights (Meas Seq=calibration measurement sequence; IFGs=number of interferograms; fw=forward sweep; bw=backward sweep; DET=detector; PREA=preamplifier; BB=blackbody; Resp=response; CH=channel). Parameters a_2 and a_3 for channel 1 referenced to a DC value of 2.086 V and for channel 2 to a DC value of 1.700 V.

Date (Flight)	Meas Seq	IFGs (fw/bw)	T_{DET} [K]	T_{PREA} [K]	T_{BB} [K]	DC [V]	a_2	a_3	ΔResp %
RECONCILE 2010, CH 1									
Jan 17th (FL01)	2	10/5	4.41	231.40	202.70	1.754	-1.702	12.702	16.7
Jan 22nd (FL03)	1	7/12	4.48	232.25	206.56	2.003	-2.058	9.313	17.8
Jan 24th (FL04)	2	13/9	4.45	232.84	209.63	1.999	-2.055	8.507	17.5
Jan 25th (FL05)	1	19/19	4.46	232.79	210.02	1.997	-2.020	7.212	16.8
Jan 30th (FL07)	3	28/27	4.48	232.57	208.17	1.857	-2.094	10.030	18.3
Feb 2nd (FL08)	2	28/28	4.52	232.73	211.05	2.070	-2.056	8.102	17.4
Feb 27th (FL09)	2	20/21	4.49	232.80	210.48	2.076	-2.039	8.135	17.3
Mar 2nd (FL10)	1	23/21	4.34	232.33	209.39	1.937	-1.978	8.175	16.9
Mar 2nd (FL11)	2	27/30	4.47	231.37	204.66	1.945	-2.063	9.888	18.0
Mar 5th (FL12)	2	19/16	4.49	232.92	210.91	2.082	-2.057	7.536	17.2
Mar 10th (PremierEX)	2b_2	10/9	4.31	232.59	209.85	1.887	-1.942	9.794	17.2
ESSenCe 2011, CH 1									
Dec 16th (FL02)	1	28/29	4.51	236.39	215.78	2.278	-1.870	5.477	15.2
ESSenCe 2011, CH 2									
Dec 16th (FL02)	1	26/23	4.50	236.43	215.71	1.768	-1.482	0.000	12.2

Furthermore, the DC value of 1.000 V corresponding to a typical zenith view spectrum is used here to estimate the change in response between spectra with high and low photon fluxes.

During RECONCILE, the individual percentages of the estimated changes in channel 1 detector responsivity scatter within less than $\pm 1.0\,\%$ around an average value of 17.4 %. Weak/moderate correlations with the preamplifier temperature and the temperature of the blackbody source might reflect an overall influence of the temperature on the instrument electronics. Significantly different parameters for channel 1 at the ESSenCe flight are a consequence of modifications in the amplifier setup.

3.1.2 Fourier Transformation of interferograms and phase correction

MIPAS-STR records double-sided interferograms of atmospheric radiation in the path domain (path coordinate s). The raw interferograms are converted into raw spectra in the wave number domain (wave number $\tilde{v}$) via Fast Fourier Transformation. In this context only the basic concept of Fourier Transformation can be introduced. More detailed discussions of FTIR spectroscopy in general are given for example by Hollas [2003] (and references therein) and specific to MIPAS instruments by Blom et al. [1996], Trieschmann [2000], Kimmig [2001] and Kleinert [2003]. The raw interferograms IFG are related to the raw spectra A via the phase function ϕ as follows:

$$A(\tilde{v}) = Re\left[e^{-i\phi(\tilde{v})} \int_{-\infty}^{\infty} IFG(s)e^{-i2\pi\tilde{v}s}ds \right] \qquad (3.3)$$

Significant radiation contributions arise from inside the instrument and partly show a different phase compared to the radiation of interest (i.e. atmospheric radiation). As a consequence, the measured spectra show a natural phase ϕ_{nat} (see Kleinert and Trieschmann [2007]). Phase errors result from sampling shifts relative to the interferogram peaks and from the frequency-dependent signal propagation in time in the optical dispersive elements and in the electronic components. The phase errors ϕ have to be corrected and must therefore be separated from ϕ_{nat}.

MIPAS-STR measures weak thermal radiation signatures from the atmosphere. Although the optics are dry-ice cooled to $\sim$210 K, the significant radiation contributions from optical components inside the instrument cannot be neglected. The real parts of the spectra basically contain the atmospheric radiation, and instrumental radiation from the telescope as well as optical components between the beamsplitter and the detector (with a phase difference of 180°). Emission of the beamsplitter mainly contributes to the imaginary parts of the spectra.

An advanced approach for the phase correction of MIPAS instruments is described by Trieschmann et al. [1999] and Kleinert and Trieschmann [2007]. The key steps are the extraction of a characteristic instrumental phase function from blackbody calibration measurements which then serves as starting point for a statistical phase correction approach for the atmospheric measurements. The individual steps are schematically summarised in Figure 3.1. In the first step of phase determination, a preliminary phase is extracted from the blackbody calibration measurements (included in each limb sequence) according to Forman et al. [1966], neglecting the comparably weak beamsplitter emission versus the source radiation of the blackbody calibration measurements. In the next step, the resulting averaged preliminary phase function is applied to the upward viewing measurements (5°/8°/10°) and the beamsplitter emission is extracted according to Trieschmann et al. [1999]. These measurements show only weak atmospheric signatures and allow for the extraction of the significant beamsplitter emission. The resulting imaginary spectra are averaged and smoothed. This reduces the noise level, whereas the broad beamsplitter emission patterns are not attenuated. In the following step, an improved instrumental phase function is extracted again from the blackbody calibration measure-

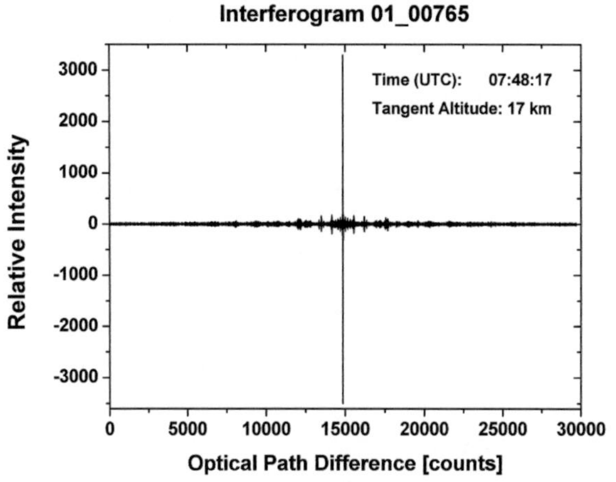

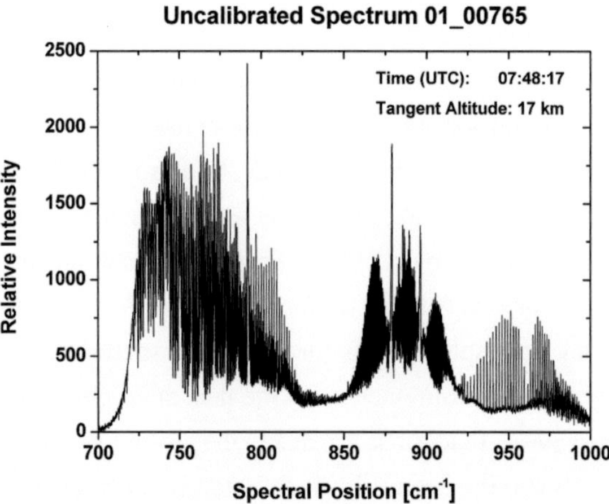

Figure 3.3: Examples for a measured atmospheric raw interferogram (upper panel) and the resulting raw spectrum (lower panel) after Fast Fourier Transformation for the RECONCILE PremierEX flight on March 10th 2010 (flight altitude 17.8 km).

ments according to Forman et al. [1966] considering the previously extracted beamsplitter emission spectrum. The resulting final averaged instrumental phase function then serves as starting point for the statistical phase correction approach for the atmospheric measurements. The iterative approach introduced by Trieschmann et al. [1999] follows two main assumptions: (i) Correlations between the real and imaginary parts of the spectra are minimised, as these parts are theoretically independent from each other. (ii) Variances in the imaginary parts of the spectra are minimised, since the imaginary spectra basically represent the solid state body emission of the beamsplitter and no sharp line signatures are expected. From Fourier Transformation and phase correction, atmospheric raw spectra are obtained and are then subject of further calibration steps. Figure 3.3 shows an exemplary atmospheric raw interferogram together with the resulting raw spectrum from Fourier Transformation and phase correction.

3.1.3 Determination of instrumental offset and radiometric calibration

The next steps within data processing towards radiometric calibrated spectra are the extraction of the remaining instrumental offset in the atmospheric raw spectra and the determination of the radiometric gain function (Figure 3.1). Taking into account the instrumental offset U and the radiometric gain function C (both fitted linearly to the measurement times), the atmospheric raw spectra A are transferred into calibrated spectra S:

$$S(\tilde{v}) = \frac{A(\tilde{v}) - U(\tilde{v})}{C(\tilde{v})} \tag{3.4}$$

The iterative scheme for the extraction of the instrumental offset from the zenith view calibration measurements towards cold space was developed by Höpfner et al. [2001b]. Beside the instrumental background radiation to be characterised, zenith view observations from an aircraft also contain weak atmospheric signatures which have to be removed accurately.

43

Therefore, a preliminary calibration is applied to the zenith view observations. The zenith view measurements are linearly extrapolated to the beginning and end of the calibration interval, yielding two characteristic zenith view spectra. In the resulting two spectra, signatures from trace gases are fitted with Gaussian functions and are then subtracted, yielding a first approximation for the instrumental offset. Taking this first approximation of the instrumental offset into account, a preliminary radiometric gain function is extracted from the blackbody calibration measurements, allowing for a preliminary calibration of the zenith view spectra (see Equation 3.5). The preliminarily calibrated zenith view spectra are then subjected to a radiative transfer line fitting step utilising the forward model KOPRA and the inversion module KOPRAFIT (see Section 3.2.2) to reproduce the observed atmospheric signatures with improved accuracy.

Retrieval parameters are the volume mixing ratios of CO_2, O_3, HNO_3 and H_2O along with the fitting parameters spectral shift, scaling and also radiometric offset, as this characteristic is not yet specified accurately in the previous step. The retrieval setup applied in this work allows rather flexible line fitting, since the scope of the retrieval is not to reproduce the mixing ratios of the trace gases accurately, but to reproduce the observed lines best utilising information on spectral and instrumental line shapes. Fitting of the complete spectrum at once is avoided, since the preliminary offset calibration shows different systematic offset errors for different regions of the spectrum. Instead, fitting is performed in individual microwindows in order to capture offset variations in different spectral regions of the spectrum.The radiative transfer step is carried out for each single zenith view measurement and yields the individual instrumental offset spectra after substraction of the retrieved spectral signatures. The individual offset spectra are then again linearly extrapolated in time to the beginning and end of the measurement phase in order to reduce the measurement noise. The resulting two characteristic spectra are then linearly interpolated in time to the blackbody calibration and atmospheric measurements for further calibra-

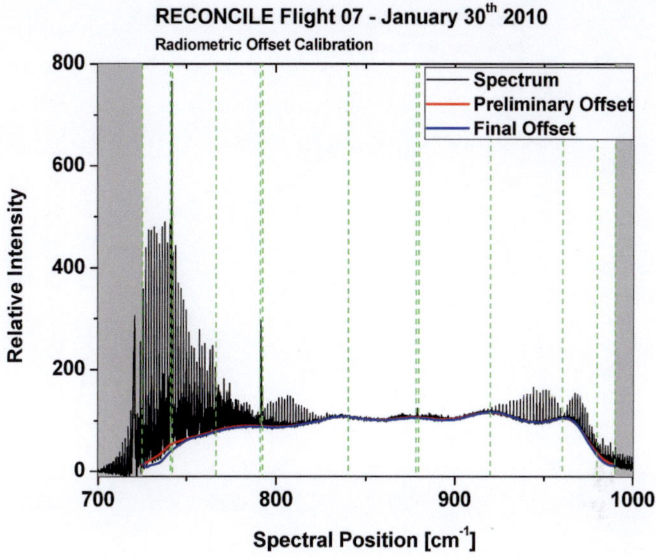

Figure 3.4: Example for a channel 1 zenith view spectrum for radiometric offset calibration, together with the corresponding preliminary and final instrumental offset spectrum after the radiative transfer step. Spectral microwindows for radiative transfer step are marked by dotted green lines. Regions outside the useful spectral range of channel 1 are shaded in grey. Flight altitude 17.8 km.

tion steps. Figure 3.4 shows an example of an extrapolated zenith view spectrum.

From this spectrum first the preliminary offset spectrum (red) is determined, allowing then for the determination of the final instrumental offset (blue) via the the radiative transfer step. The spectral microwindows selected for the radiative transfer step are also indicated.

Regions containing strong Q-Branches are treated separately, since the retrieval of the Q-branches together with the associated P- and R-branches in the same microwindow can lead to significant residuals as the conse-

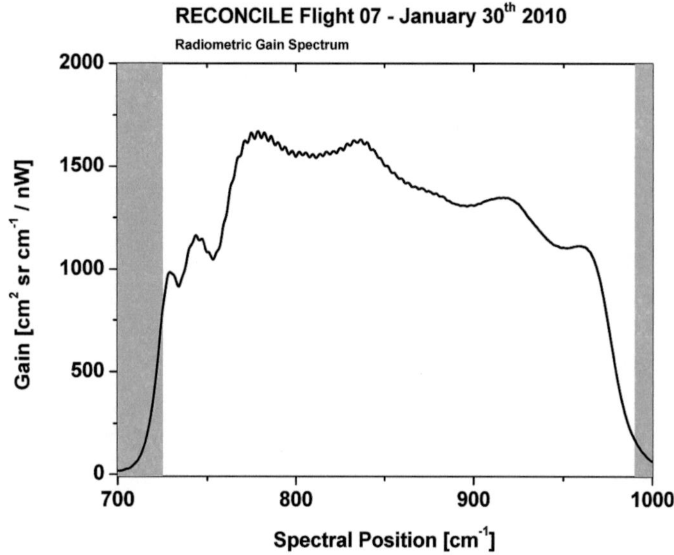

Figure 3.5: Example for a channel 1 radiometric gain spectrum for radiometric calibration. Oscillations on the scale of a few wave numbers result from Fabry-Perot channelling effects on the Si:As detector substrate. Regions outside the useful spectral range of channel 1 are shaded in grey.

quence of improper representation of line-mixing effects. The improvement in instrumental offset reconstruction by the radiative transfer step compared to the preliminary offset correction can be seen especially in the region with $\tilde{\nu} < 800$ cm⁻¹ which is densely populated with spectral lines.

Taking into account the instrumental offset spectrum U, the radiometric gain function C is determined from the blackbody measurements BB taking into account the Planck Function P of the blackbody with the measured temperature T and the emissivity em:

$$C(\tilde{\nu}) = \frac{BB(\tilde{\nu}) - U(\tilde{\nu})}{P(\tilde{\nu}, T) \cdot em(\tilde{\nu})} \qquad (3.5)$$

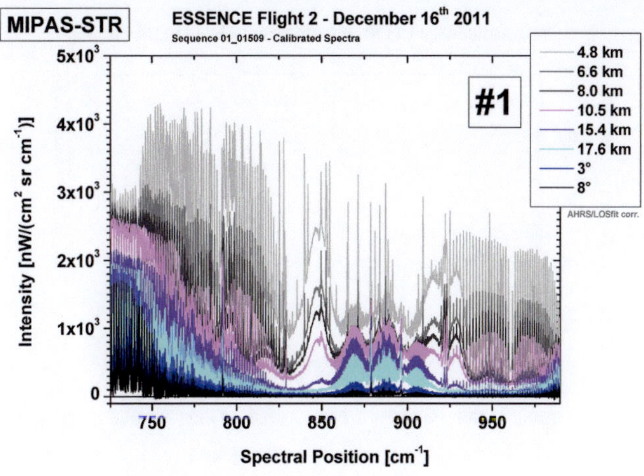

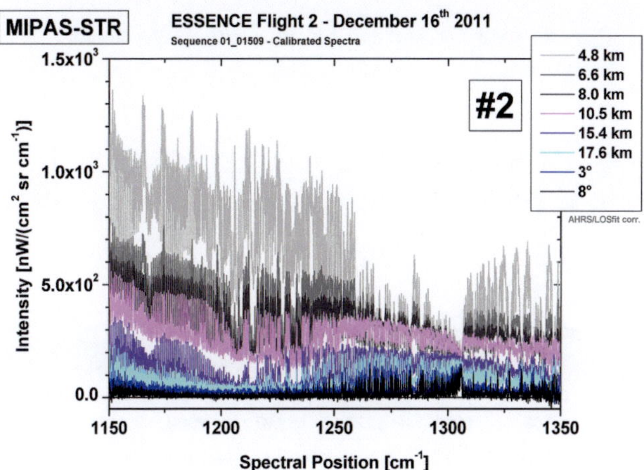

Figure 3.6: Examples for calibrated channel 1 and channel 2 spectra. Legend: Tangent altitudes of limb-viewing observations and elevation angles of upward-viewing observations (0° corresponding to horizontal view). Flight altitude 17.7 km.

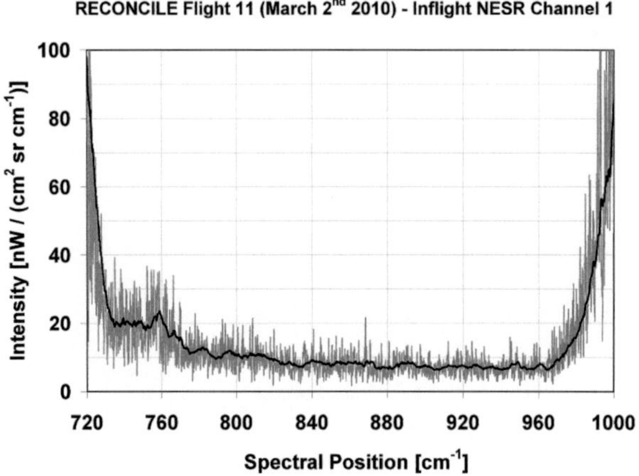

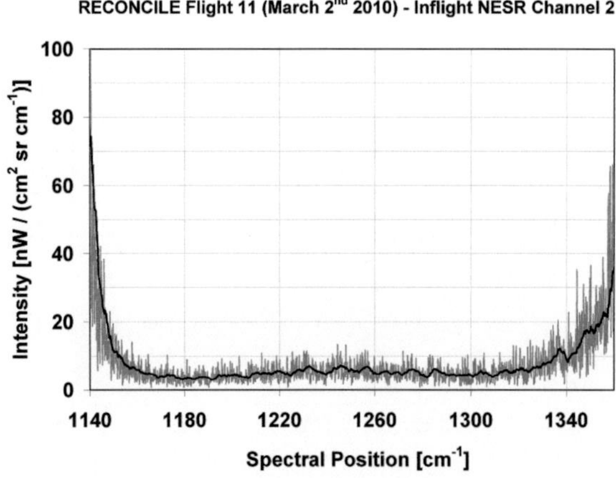

Figure 3.7: Characterisation of NESR of apodised MIPAS-STR channel 1 (from Woiwode et al. [2012]) and 2 spectra. NESR values at individual spectral datapoints (grey) and moving average (black).

As in the case of the instrumental offset spectra, the radiometric gain spectra are determined from the individual blackbody measurements and then extrapolated linearly to yield two characteristic spectra with reduced noise for the begin and end of the calibration phase. These two spectra again allow linear assignment of the radiometric gain in time to the atmospheric measurements. In Figure 3.5 an exemplary radiometric gain spectrum is shown. The radiometric gain spectrum is characterised by patterns from the transmitting optical components (beamsplitter, ZnSe entrance window of detector dewar, optical bandpass-filter) and the frequency-dependent amplification by the electronics and reflects the shape blackbody calibration spectra.

Examples of finally calibrated channel 1 and 2 spectra are shown in Figure 3.6. The spectra from high elevation angles show no significant radiometric offset and reflect the high quality of the offset calibration. Spectra from lower elevation angles show characteristic continuum-like offsets resulting from aerosols and overlapping broad spectral signatures. Such signatures are typical for measurements in the lower UTLS region and have to be considered in retrievals.

The noise in the final calibrated spectra is quantified by the noise equivalent spectral radiance (NESR), which is shown in Figure 3.7 for typical individual in-flight measurements. Minimum NESR values below 10 $nW/(cm^2$ sr $cm^{-1})$ are obtained for channel 1 and below as 5 $nW/(cm^2$ sr $cm^{-1})$ for channel 2. This corresponds to signal-to-noise ratios of up to more than 100 and allows for retrievals of atmospheric constituents with low mixing ratios and weak spectral signatures.

3.1.4 Characterisation of the vertical field-of-view weighting function

The weighting of radiation entering the FOV under different vertical angles by the collecting optics (Winston cones) of the detector is essential for retrievals. Inhomogeneous weighting of radiation from different altitudes entering the FOV during individual measurements can result in systematic altitude offsets in retrievals and affect the absolute retrieved values.

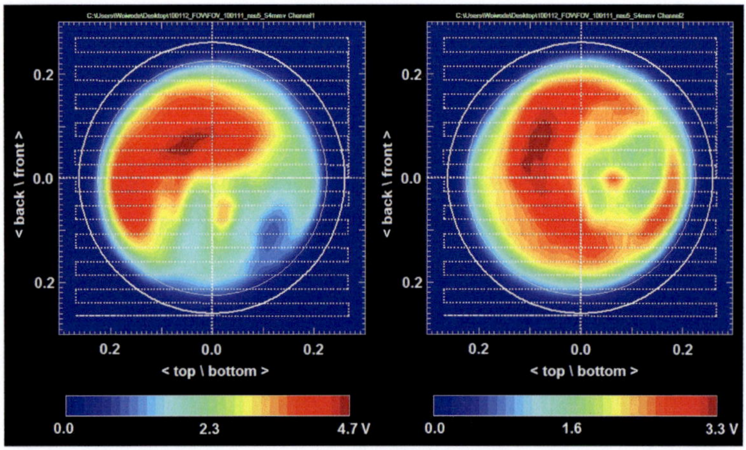

Figure 3.8: Measured FOV patterns of spectral channels 1 (left) and 2 (right) on January 11th 2010. Dotted white lines indicating positions of blackbody source during measurement. Colour coding representing interpolated DC measured at the detector of the corresponding channel. x/y-Axis corresponding to angles in degrees.

For comparison, the vertical diameter of the MIPAS-STR FOV is about 3 km at the lowest tangent point at $\sim$5 km altitude for a flight altitude of 18 km, while vertical structures in the order of 1-2 km are resolved in the retrieval.

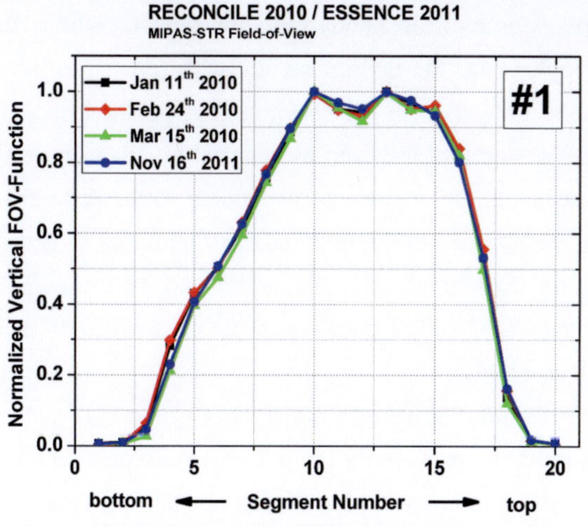

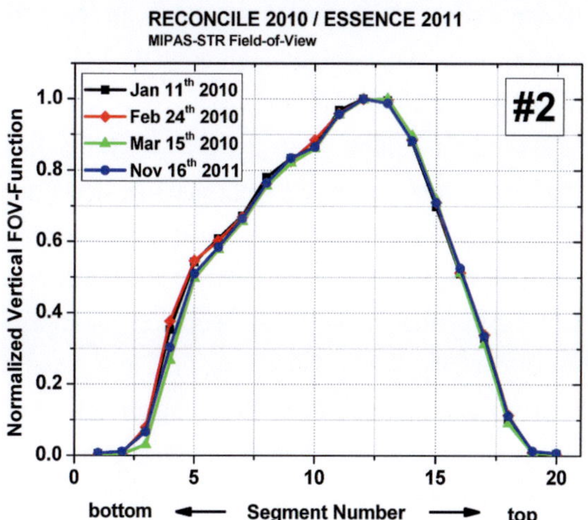

Figure 3.9: Characterisation of relative vertical FOV functions for MIPAS-STR channels 1 and 2.

The MIPAS-STR FOV is characterised by calibration measurements on ground. A spot source of blackbody radiation (250°C) with a diameter of 0.1 mm is coupled into the instrument under different positions and angles and thereby imaged onto the detector. The position of the source is varied utilising a step-motor-driven moving table. To visualise the FOV pattern, the measured DC values corresponding to the different positions and angles of the source are interpolated. In Figure 3.8 typical measured FOV patterns for the MIPAS-STR channels 1 and 2 are shown. The non-symmetric shapes of the FOV patterns are features of the Winston Cones inside the detector dewar.

In the retrievals only the vertical component of the FOV pattern is considered. Therefore, the measured FOV patterns are separated into 20 slices along the top-bottom-axis and are each integrated to obtain the vertical FOV weighting function. In the retrieval setup, the relative FOV weighting function with unitless x-axis is implemented together with the FOV opening angle to characterise the vertical weighting through the FOV (the full-cone diameter of the MIPAS-STR FOV is $0.44°$).

Figure 3.9 shows the resulting vertical FOV weighting functions for MIPAS-STR channels 1 and 2 from different measurements during REC-ONCILE and ESSenCe. The individual measured functions remain practically constant over a total period of approximately 2 years, covering several cooling cycles of the instrument and a non-operating phase of ~1.5 years between the two campaigns. Taking into account the low overall sensitivity of the retrieval on minor changes in the FOV (compare Stiller et al. [2002]) and the high reproducibility of the FOV measurements, no significant errors due the FOV characterisation are expected for the retrievals discussed in this work.

3.1.5 Post-processing and characterisation of pointing information

Active pointing stabilisation and accurate knowledge of the LOS elevation angle and the observer altitude are essential for accurate retrievals from the spectra. The MIPAS-STR pointing system is described by Keim [2002] in detail. Attitude information is recorded at 128 Hz by an inertial navigation system (Attitude and Heading Reference System, AHRS) installed in a fixed relation to the scanning optics. A fast active scan mirror control loop including an electric motor and an 19-bit encoder is used for active pointing stabilisation and to compensate attitude changes of the flying platform. For a flight altitude of 18 km a systematic vertical pointing error of 2.5 arcmin results in an vertical pointing offset of less than 0.03 km at the highest tangent point (17.9 km). In contrast, for the lowest tangent point at 5 km altitude an offset of about 0.3 km is resulting, leading to systematic errors in retrievals. In this work, the pointing knowledge was optimised and errors were characterised to estimate the associated uncertainties in retrievals.

The quality of the final pointing information is affected by the following uncertainties: (i) The accuracy of the scan mirror control, (ii) the quality of the reference attitude information from the navigation system and (iii) thermal warping between the navigation system (mounted on top of instrument) and the dry-ice cooled optical components both directly in contact with stratospheric air during flight.

The accuracy of the scan-mirror control is characterised by comparison of the actually measured angle of the scan-mirror and the commanded angle of the scan mirror at a certain time. The commanded scan-mirror angles are extrapolated from the attitude and its variation measured by the AHRS into the future, with a latency of the considered AHRS data versus the commanded scan-mirror angle of $\sim$11.4 ms. Figure 3.10 shows the commanded scan-mirror elevation angles (in the encoder reference system) and the actually measured encoder elevation angles for a smooth flight period (left

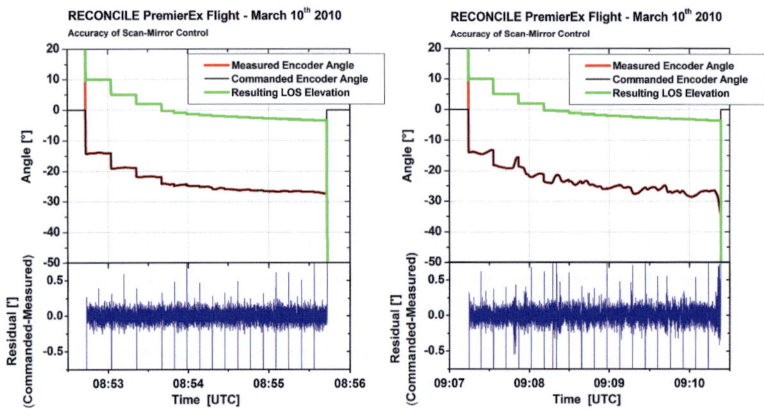

Figure 3.10: Upper panels: Comparison of measured and commanded scan mirror angle (encoder coordinate system) together with resulting LOS elevation angle (horizontal coordinate system, $0°$ elevation = horizontal view) in degrees. Lower panels: Residual between commanded and measured encoder angle in arcmins.

panel) and a perturbed flight period with roll-variations of the aircraft in the order of several degrees (right panel).

The graphs of the commanded and measured encoder angles reflect how the scan mirror control compensates the roll variations of the aircraft. Also shown are the resulting effective elevation angles of the instrument with reference to the horizon (horizontal view=$0°$), proving that roll-variations of the aircraft are compensated to a high degree. The step-like pattern results from the observation angles stabilised subsequently during a limb scan. At the begin and the end of the shown limb scans (high elevation angles towards low elevation angles) zenith view calibration measurements (measured encoder angle $> 0°$) and blackbody calibration measurements (measured encoder angle $< -30°$) were carried out. Here, the 'commanded encoder angle' is not representative. The residuals between the commanded and measured scan-mirror elevation angles in the encoder reference system shown in the lower panels of the plots indicate that the scan mirror control reliably follows the commanded angles. 'Spikes' result from changes in the commanded limb-viewing angles by the automatic sequence control and from fast attitude variations of the flying platform. A typical residual of ± 0.20 arcmins is found for the unperturbed flight phase and of ± 0.25 arcmins for the perturbed flight phase. No significant systematic offset is identified. Considering the shown residuals, the effective net contribution to the LOS uncertainty by the scan mirror control is estimated to 0.09 arcmin (1σ) for individual limb observations (sampling time ~ 9.5 sec).

It has to be mentioned that for rapid and steep turns (i.e. between two flight legs) the scan mirror control loop may not follow the changes in inclination sufficiently fast. Therefore, during steep turns additional uncertainties might result from the scan mirror control loop. Furthermore, under such conditions the quality of the attitude data from the AHRS may by degraded. Therefore, aircraft turns between different flight legs are marked in the retrieved vertical cross-sections in Section 4 to indicate the possibility of enhanced retrieval errors.

The estimated precision of the AHRS pitch angles (AHRS pitch axis approximately coinciding with the axis of the MIPAS-STR pointing mirror and the roll axis of the aircraft) given by the supplier SEG mbH, Riegel (Germany) is 0.5 arcmin (1σ) under quiet conditions. Since the MIPAS-STR AHRS is operated on a dynamic airborne platform characterised by vibrations the quality of the AHRS pitch data is lower. To improve the quality of the AHRS pitch information, the data are post-processed.

The AHRS is an inertial navigation system making use of the combination of fixed gyros and accelerometers, calculating its actual alignment versus a virtual horizontal platform (see Maucher [1999] and Keim [2002]). Platform errors of the AHRS are estimated based on a real-time Kalman-Filtering approach, giving a non-scaled parameter suitable for the post-flight correction of drifts in the AHRS data at a timescale of minutes. The

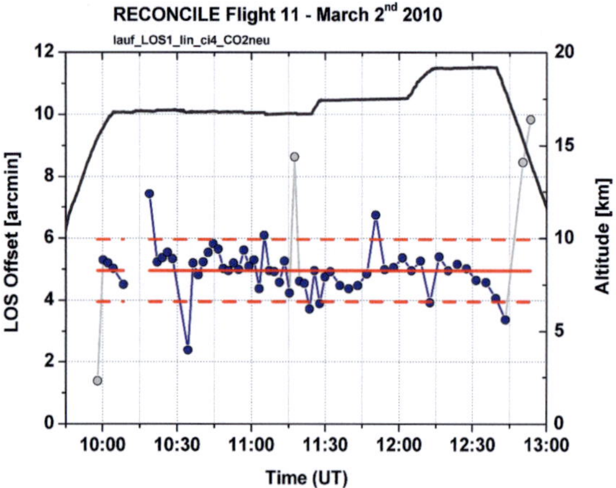

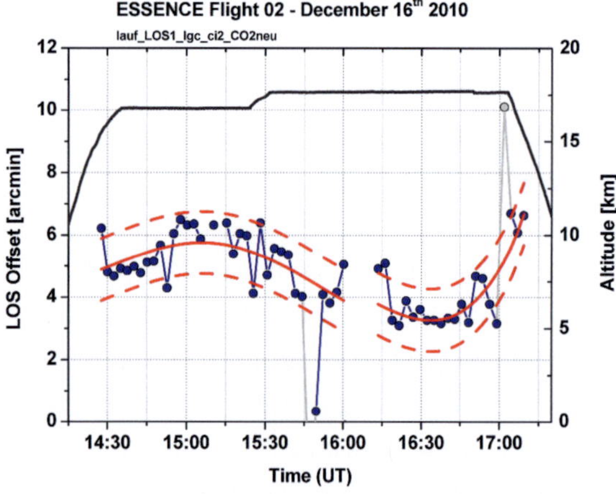

Figure 3.11: Characterisation of systematic LOS offset by retrieval. Blue dots: Retrieval results for individual sequences; grey dots: excluded values. Dark grey line: flight altitude. Red solid and dotted lines: LOS offset correction and adopted uncertainty range (Upper panel: from Woiwode et al. [2012] with modifications).

suitability of this parameter for post-flight refinement of the elevation angles was found empirically by Maucher [2009] for the comparable inertial navigation system of the ballon-borne MIPAS-B2 instrument and was referenced to the highly precise pointing information obtained from the MIPAS-B2 star-reference system (Maucher [1999]).

For MIPAS-STR, this initially non-scaled parameter resulting from Kalman filtering is also applied for post-flight pointing refinement and is calibrated using AHRS pitch measurements under quiet conditions on ground with the aircraft being untouched. Under these conditions, measured virtual pitch variations are attributed to errors and are echoed into the non-scaled correction parameter, which then can be scaled accordingly. For different flights the scaling factor is reproducibly found to be $\sim$15. Sensitivity considerations result in an estimated uncertainty of 0.75 arcmin (1σ) for the AHRS pitch data under flight conditions after post-processing (compared to an estimated uncertainty of 2 arcmin (1σ) prior to post-processing).

Systematic errors due to thermal warping of the instruments optical components relative to the reference system of the AHRS are quantified via LOS retrievals (for retrieval details see chapter 3.2.2). In Figure 3.11 the results from the LOS retrieval for the RECONCILE Flight 11 on March 2nd 2010 and the ESSenCe Flight 02 on December 16th 2011 are shown together with the LOS offset correction applied to the following retrievals.

For the flight on March 2nd 2010, the retrieved LOS offset values are scattered around a mean LOS offset of $\sim$5 arcmin and can be approximated by a constant offset correction. Only for the last scans a significant trend towards lower offset values is found, which is still in the uncertainty range adapted for error estimation in the following retrievals. The flight on December 16th 2011 took place under rather cold conditions. Systematic variations on timescales of several minutes reflect variable warping of the instrument probably due to thermal drifts. Here, a fourth-order polynomial fit resembles the trends in the retrieved LOS offset values reasonably.

For the ESSenCe and RECONCILE flights systematic LOS errors due to thermal warping in the order of 5 arcmin were found and corrected by LOS-retrievals.[2]. For the LOS-correction obtained from retrieval an uncertainty of 1 arcmin (1σ) is estimated (indicated by dashed red lines in Figure 3.11). The total estimated pointing error is obtained via combination of the error resulting from the scan-mirror control (net 0.09 arcmin for a complete interferometer sweep), the accuracy of the AHRS pitch data (0.75 arcmin) and the conservative estimation of the accuracy of the retrieved systematic LOS offset attributed to thermal warping (1 arcmin), yielding a total LOS-error[3] of 1.25 arcmin (1σ).

3.1.6 Summary: Optimisation and characterisation of level-1 product

To provide the basis for accurate retrievals from the MIPAS-STR measurements, the level-1 processing chain was optimised and standardised for the spectral channel 1 and the measurements were comprehensively characterised. For the first time, also channel 2 spectra from MIPAS-STR were fully calibrated and characterised. The final level-1 product (calibrated spectra with with known geolocation and FOV information) is characterised as follows:

- Radiometric offset calibration: A new set of microwindows was exploited and a flexible fitting setup was chosen, allowing further reduction of residuals in the radiative transfer step.

- Detector nonlinearity: Changes in detector response through nonlinearity were characterised for channel 1 for 11 flights during REC-

---

[2] The slightly different LOS correction discussed in Woiwode et al. [2012] is valid in context of an outdated FOV implementation.

[3] The smaller LOS-error of 0.78 arcmin described in Woiwode et al. [2012] considers for the retrieved LOS-offset only the scattering of the individual data points (0.19 arcmin) around the average value. To consider dynamic systematic variations during flight and potential further uncertainties from retrieval, here this more conservative estimate is applied.

ONCILE and one ESSenCe flight. Furthermore, the detector non-linearity of channel 2 was characterised exemplaric for the ESSenCe flight. During RECONCILE the determined changes in detector response between spectra with low and high photon fluxes were found to vary within less than $\pm 1\%$ around a mean value of 17.4 % for channel 1.

- FOV characterisation: The MIPAS-STR FOV was characterised for the RECONCILE and ESSenCe campaings. The vertical FOV functions derived from characterisation measurements within a period of approximately 2 years covering several cooling cycles of the instrument were found to vary insignificantly.

- Pointing knowledge: The accuracy of the scan-mirror control was verified by analysing the consistency between commanded and measured pointing mirror elevation angles for quiet and disturbed flight sections. Furthermore, the post-flight attitude information of the AHRS pitch used for LOS stabilisation was refined by an additional post-processing step considering error information recorded inflight by the Kalman filtering procedure. Remaining estimated uncertainties of the AHRS pitch values used for pointing stabilisation were reduced from 2 arcmin (1σ) to 0.75 arcmin (1σ) under flight conditions. An additional step for correction of systematic pointing errors through thermal warping of the optical components was included in data processing. Approximately constant or slowly varying systematic pointing offsets versus time were quantified by a separate retrieval step. Typical systematic pointing corrections were reproducibly determined in the order of 5 arcmins for different flights.

- The final level-1 product is characterised as follows: (i) Radiometric accuracy: Considering the uncertainties resulting from radiometric offset calibration, radiometric gain calibration and detector non-linearity, a combined (random+systematic) radiometric error of 2 %

(1σ) is estimated for measured radiances in the order of 10^3 nW/(cm^2 sr cm^{-1}). Typical NESR values for channel 1 (channel 2) spectra are found to be in the order of 10 (8) nW/(cm^2 sr cm^{-1}). (ii) Postflight pointing knowledge: The knowledge of the post-processed vertical pointing information is estimated to 1.25 arcmin (1σ) for individual limb observations.

3.2 Retrieval (level-2 processing) and validation

This section discusses the extraction of atmospheric parameters from the calibrated spectra and the characterisation and validation of the retrieval results. It starts with the discussion on the influence of clouds and aerosol on the measurements and their detection. Cloud and aerosol signatures are characteristic for FTIR limb measurements in the UTLS region and set requirements for the retrieval setup. Then, after a brief introduction into the retrieval method, the retrieval setup and strategy is elaborated. Finally, the retrieval results are characterised and validatated with collocated in-situ measurements.

3.2.1 Cloud detection

Clouds and aerosol particles give rise to greybody-like signatures in thermal limb-emission spectra. Depending on the particle size, abundance and composition, clouds can be partially transparent (i.e. cirrus clouds and diffuse PSCs) or optically thick (i.e. dense tropospheric water vapour/ice clouds), hindering infrared limb-emission observations of the atmospheric layers behind and below. Information on cloud-coverage can be derived directly from limb-emission spectra by applying the cloud-index introduced by Spang et al. [2004]. Accordingly, the colour ratio between the spectral microwindows 788.20 to 796.25 cm^{-1} and 832.30 to 834.40 cm^{-1} indicates clouds for values close to 1, whereas values higher than 4 indicate cloud-free conditions. Values in between indicate partially cloud affected spectra (i.e. in the presence of diffuse clouds or when only parts of the FOV touch cloud-affected atmospheric layers). Figure 3.12 shows examples for cloud-affected MIPAS-STR spectra: The upward-viewing spectrum (8°) indicates already a continuum-like offset indicating particles above the flight altitude[4], as these measurements were performed from inside a diffuse PSC

---

[4]Typically the baseline of upward-viewing spectra and of spectra with high tangent altitudes is close to zero under cloud-free conditions, compare Figure 3.6.

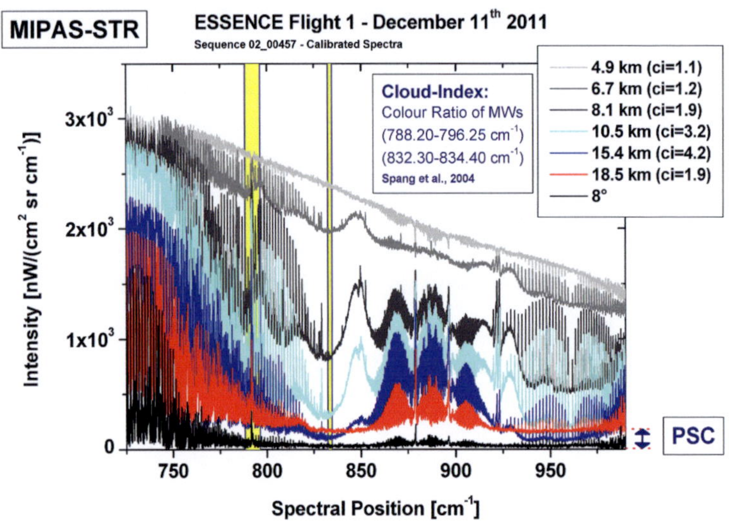

Figure 3.12: Examples for cloud-affected MIPAS-STR spectra and cloud index (ci). Legend: Tangent altitudes of limb-viewing observations and elevation angle for upward-viewing observation ($0°$ corresponding to horizontal view).

cloud. The spectrum with the highest tangent altitude (18.5 km) shows a strong continuum-like offset since the instrument was pointing approximately horizontally into a PSC layer. The spectrum corresponding to the lower tangent altitude of 15.4 km again shows a weaker continuum-offset indicating less cloud/aerosol-affected conditions there, but is also affected by the PSC layer above.

Towards the troposphere the spectra again show an increasing continuum-offset due to thin clouds or aerosol and overlapping signatures from trace gases. The spectra with tangent altitudes of 6.7 and 4.9 km are strongly cloud-affected, showing a greybody-like shape with only weak spectral lines from trace gases. The spectrum with the tangent altitude of 4.9 km is affected by an optically dense cloud at warmer tropospheric altitudes, which is indicated by a greybody-like spectrum superimposed by absorption signatures from trace gases at higher and colder atmospheric layers.

For strict cloud-filtering of the spectra recorded by the space-borne instrument MIPAS-ENVISAT, a lower cloud-index threshold of 4 is recommended. For the infrared limb sounder CRISTA partially cloud-affected spectra with cloud-index values of 2 to 4 were shown to be retrievable (Spang et al. [2004]). In the following in will be demonstrated, that retrievals from stratospheric MIPAS-STR spectra with cloud-index values down to 1.6 are possible if the retrieval is modified accordingly. This allows vertically resolved retrievals from spectra recorded inside diffuse PSC clouds and thereby a detailed reconstruction of the chemical composition of the gas phase within and below PSCs, which is hardly accessible for comparable satellite- and balloon-borne techniques.

It is mentioned that the MIPAS-STR measurements from inside PSCs during RECONCILE and ESSenCe also contain signatures characteristic for certain PSC particles. A potential 'red-shifted NAT-signature' is observed, differing from the signature of this particle species discussed by Höpfner et al. [2006]. This signature can be identified in Figure 3.12 in the spectrum corresponding to the tangent altitude of 18.5 km in the 820 cm^{-1} region. A local maximum with a 'shoulder' towards lower wave numbers is observed. These particle signatures are not exploited here, but might be subject of further studies with the MIPAS-STR data.

3.2.2 Retrieval method

The retrieval processing encompasses the reconstruction of atmospheric parameter profiles and instrumental parameters from the measured spectra. The retrieval makes use of two essential modules: The forward model allowing for the calculation of atmospheric spectra under given conditions (pressure, temperature, trace gas vmr, instrumental parameters, observation geometry) and the inversion module, allowing for fitting of the forward modelled spectra to the measurements through variation of the specific target parameters. In case of convergence and agreement between modelled and measured spectra (i.e. low residuals), the retrieved parameters associated to the forward calculation are expected to reproduce the atmospheric state or instrumental parameters best. However, error budgets as well as smoothing effects of the retrieval have to be assessed carefully when interpreting the retrieval results quantitatively. Furthermore, care has to be taken as several retrieval parameters are not independent from each other.

Within this work, atmospheric and instrumental parameters were retrieved from MIPAS-STR spectra utilising the forward model Karlsruhe Optimized and Precise Radiative transfer Algorithm (KOPRA) (Stiller [2000]) and the associated inversion module KOPRAFIT (Höpfner et al. [2001a]).

The forward model KOPRA allows numerical solving of the radiative transfer equation and yields forward spectra associated to the measurements and the analytical derivatives (spectral radiance vs. retrieval parameter) needed for the inversion process. Fast line-by-line calculations of atmospheric spectra are performed under consideration of atmospheric state parameters (vertical distribution of pressure/temperature/trace gas volume mixing ratio (vmr)) including also refraction, aerosol effects (extinction and/or scattering) and instrumental parameters (i.e. FOV and ILS (instrumental line shape)). Spectral signatures of trace gases are modelled line-by-line considering the Voigt line-shape model and spectral line data are taken from spectral databases such as HITRAN (HIgh resolution TRANsmission, i.e. Rothman et al. [2009]). For MIPAS-STR data processing, a dedicated compilation of HITRAN data for MIPAS data processing (Flaud et al. [2002]; Flaud et al. [2006]) is utilised. In cases where no accurate line data is available absorption cross-section data is applied.

KOPRAFIT utilises the analytical derivatives provided by KOPRA and allows fitting of the full set of observations of one limb scan simultaneously. Thereby, the targeted atmospheric parameter profiles and/or instrumental parameters are reconstructed. Atmospheric parameter profiles are inverted by Gauss-Newton iterations subjected to first-order Tikhonov-Phillips regularisation (Tikhonov [1963]; Phillips [1962]):

$$x_{i+1} = x_i + \left(\mathbf{K}_i^T \mathbf{S}_y^{-1} \mathbf{K}_i + \gamma \mathbf{L}^T \mathbf{L} \right)^{-1} \left[\mathbf{K}_i^T \mathbf{S}_y^{-1} \left(y - f(x_i) \right) + \gamma \mathbf{L}^T \mathbf{L} (x_a - x_i) \right] \quad (3.6)$$

with x_i representing the vector with the retrieval quantities (i.e. atmospheric parameter profiles), i the iteration index, x_a the a-priori profile, y the vector with measured radiances, $\mathbf{K}_i$ the spectral derivatives matrix (Jacobian), $\mathbf{S}_y$ the variance-covariance matrix of the measurements (i.e. including the measurement noise) and f the forward spectrum. $\mathbf{L}$ represents the first-derivative regularisation matrix. The regularisation strength is adjusted via the regularisation parameter γ.

The Tikhonov-Phillips side constraint is applied since (i) the applied retrieval grid (0.5 km) is mostly finer than the measurement grid (typically 1.0 km) and (ii) depending on actual trace gas abundances and measurement noise the inversion problem can be further under-determined. The fine retrieval grid is chosen as local vertical resolutions of ≤ 1 km are feasible due to increased sampling density around flight altitude, oversampling and additional information included in pressure-broadened spectral lines, provided the signal-to-noise ratio is sufficiently high.

The applied first-derivative constraint forces the retrieved profile in case of low information included in the measurements only to the shape of the a-priori profile but not the absolute values of the a-priori. Therefore, the Tikhonov-Phillips constraint avoides biases in the retrieved profiles towards the a-priori (Steck [2002]), which is especially important for statistical studies with retrieval results. Regularisation was applied to the retrieval parameters yielding vertical profiles (i.e. temperature, trace gas volume mixing ratios and aerosol extinction). The regularisation strength was adjusted for each parameter and flight individually, providing that oscillations in the resulting profiles were avoided and the residuals between measured and retrieved spectra were minimised.

3.2.3 Retrieval strategy

One key issue within this work was the elaboration of a retrieval strategy tailored to cloud- and aerosol-affected observations from the UTLS region. The retrieval strategy is described in Woiwode et al. [2012] for observations only weakly affected by aerosol and is extended here also to PSC-affected observations. In contrast to its balloon-borne and satellite-borne versions, MIPAS-STR completely focusses on the altitude region between $\sim$5 and 20 km with dense vertical and horizontal sampling, an altitude region which is particularly challenging for the retrievals. As the goal was to resolve mesoscale chemical and dynamical structures with vertical extensions in the order of 1 km, a retrieval setup allowing the extraction of maximal information from the spectra was required. Compared to processing of spectra from higher atmospheric layers ($>$20 km), several effects complicate retrievals from infrared limb-emission measurements from the lower UTLS region:

- Low number density particles and aerosol as well as thin cloud layers (i.e cirrus) under nominal cloud-free conditions

- Broad spectral signatures from complex molecules not known or not considered in the retrieval processing

- Spectral interference with significantly pressure-broadened line wings of adjacent spectral signatures and line-shape-related effects

Although the weights of these contributions are variable for different spectral regions, these effects lead to approximately continuum-like contributions within the range of the spectral microwindows used for the retrievals here (typically a few cm^{-1}). The applied retrieval strategy is based on the following guidelines:

- The effects resulting in continuum-like contributions on the scale of retrieval microwindows are treated by the retrieval of wave number independent continuum extinction. In case of strong continua (i.e. in PSC clouds), continuum extinction is inverted logarithmically to meet the high dynamics in the continuum background between the different measurement geometries.

- The simultaneous reconstruction of many different retrieval parameters (especially trace gases) is avoided, keeping the total number of retrieval parameters for each retrieval run as low as possible. In the context of the complications listed above, certain retrieval parameters can compensate each other partially in a multi-target retrieval. The mandatory simultaneous inversion of spectral shift can further amplify this drawback by directing the retrieval towards more erroneous results potentially characterised by smaller residuals.

- The retrieval is carried out following a tree-like approach, starting with the determination of a LOS correction and followed by the retrievals of temperature and trace gases. Trace gases only weakly affected by spectral interference with other signatures are retrieved first, and the resulting profiles are kept fixed for subsequent retrievals of species with signatures affected by these gases. Thereby, the aim is to achieve a consistent set of retrieved spectra corresponding to a consistent set of associated retrieved profiles.

In the following, the particular retrievals of the LOS-correction, temperature and trace gas mixing ratios are discussed in more detail. The presented retrieval scheme has been successfully applied for measurements with tangent altitudes down to $\sim$5 km under Arctic winter conditions. The spectral microwindows selected for the retrievals are listed in Table 3.2. Key requirements for the selection of microwindows were low spectral interference of the target signature with other spectral signatures and sufficiently

Table 3.2: Microwindows and target signatures selected for the retrievals (from Woiwode et al. [2012]). Central spectral positions of the most prominent target signatures are listed. For cross-section data, the corresponding broad and unresolved signatures cover the whole ranges of the microwindows. Spectral line and cross-section data were taken from the MIPAS database (Flaud et al. [2002]; Flaud et al. [2006]) and Wagner and Birk [2003].

Retrieval target	Microwindow $[\mathrm{cm}^{-1}]$	Target signature	Spectral position $[\mathrm{cm}^{-1}]$
LOS/temperature	810.1–813.1	R24e $(11101 \leftarrow 10002)$	810.93
$(CO_2$ signatures)		R26e $(11101 \leftarrow 10002)$	812.48
	955.6–958.5	P6e $(00011 \leftarrow 10001)$	956.19
		P4e $(00011 \leftarrow 10001)$	957.80
HNO_3	866.0–870.0	$v5/2v9$ bands	9 resolved net peaks ($v5$-band)
O_3	780.6–781.7	$v2$-band	780.80
			781.13
			781.18
			781.52
	787.0–788.0	$v2$-band	787.13
			787.46
			787.86
CFC-11	842.5–848.0	$v4$-band	cross-section
CFC-12	918.9–920.6	$v6$-band	cross-section
	921.0–922.8	$v6$-band	cross-section
$ClONO_2$	779.8–780.5	$v4$-band	cross-section
	805.1–805.5	$v3$-band	cross-section
H_2O	795.7–796.1	pure rotation	795.89

strong opacity of the target signatures without being saturated. Furthermore, spectral interference with H_2O signatures was avoided, as these signatures strongly increase in the troposphere and concentrations are highly variable in these altitudes.

Within each microwindow, continuum extinction and spectral shift of the microwindow were also retrieved beside the target parameter. An important precondition for the accuracy of the retrievals was a reliable offset calibration of the spectra (Section 3.1.3), as the parameters radiometric offset and continuum extinction can correlate strongly with each other, giving rise to significant retrieval errors. Spectral shift was also reconstructed due to the limited accuracy of the spectral calibration.

Signatures of weakly interfering gases (slightly exceeding the measurements noise) were modelled with climatological profiles (Remedios et al. [2007]). The only exception was the CO_2-profile. As a precise knowledge of the vertical distribution of CO_2 is crucial for the LOS and the temperature retrievals, the CO_2 profile was constructed from highly accurate in-situ measurements[5] during RECONCILE and under consideration of trends in atmospheric CO_2 (Mauna Loa record, http://www.esrl.noaa.gov/gmd/ ccgg/trends/) and kept fixed. As the vertical profile of CO_2 is well known, the spectral signatures of CO_2 are suited best for the reconstruction of the LOS offset and for temperature retrievals.

Pressure profiles for the retrievals were interpolated in space and time from the ECMWF T106 grid-point analysis (see http://www.ecmwf.int/) provided by NILU (Norwegian Institute for Air Research, see http://www.ni-lu.no/). Initial guess profiles (i.e. starting profile for the first retrieval iteration) and a-priori profiles (with their first derivative used for the retrieval constraint) were identical profiles in the retrievals discussed in this work. For the LOS and temperature retrievals, the initial guess/a-priori profiles for temperature were also interpolated from the ECMWF T106 grid-point analysis. For trace gas retrievals climatological profiles from Remedios et al. [2007] were used as a-priori/initial guess.

The 'retrieval tree' applied for the subsequent reconstruction of the LOS correction, temperature and the trace gases discussed in this work is indi-

---

[5]HAGAR gas chromatograph, Courtesy Michael Volk, Department of Physics, University of Wuppertal, Germany.

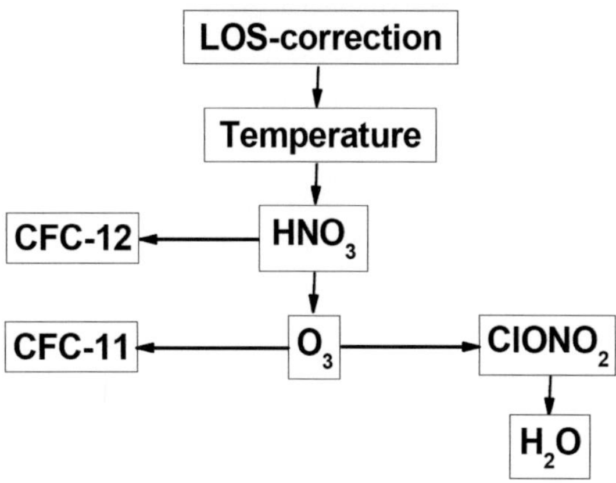

Figure 3.13: 'Retrieval-tree 'for the reconstruction of LOS correction, temperature and the indicated trace gases (from Woiwode et al. [2012]).

cated in Figure 3.13. For example, the parameters LOS correction, temperature and HNO_3 were retrieved prior to the O_3 retrieval and the resulting profiles where kept fixed in the following retrievals.

The first step was the reconstrucion of the LOS correction using spectral lines of CO_2, whereas for each limb scan one individual offset parameter was determined. Depending on the stability of the retrieved LOS offset over a flight, either the average LOS offset of all profiles of a flight or a polynomial fit was applied as LOS correction (see Section 3.1.5 and Figure 3.11). The reconstruction of the LOS offset is based on the following assumptions:

- Temperature and pressure profiles interpolated from the ECMWF T106 grid-point analysis are sufficiently accurate on average for flight sections/the entire flights. This assumption is supported by the high quality of the ECMWF data and the fact that many profiles along extended flight tracks are considered.

- The used CO_2 profile and spectral line data of CO_2 (accuracy between 2 and 5 %, see Flaud et al. [2002] and references therein) are sufficiently accurate. This is supported by the high quality of the collocated in-situ measurements used for construction of the CO_2 profile and the fact that different spectral lines of CO_2 from two different bands are used. Minor systematic LOS errors resulting from uncertainties in the spectral line-data cannot be excluded, but are expected to be within the overall estimated LOS uncertainty.

- Further error sources (i.e. relative LOS information between single measurements or radiometric errors) show no significant systematic component. This is supported by the fact that the individual retrieved LOS-offset values are constant within the scattering or show only slowly varying trends. The scatter of the individual retrieved LOS-offset values from adjacent limb sequences hints on systematic error components for individual limb sequences. However, using the average of the retrieved LOS-offset values or a polynomial fit in case of variable characteristic trends compensates for systematic errors of the individual LOS-offset values.

The temperature retrieval was carried out using the same microwindows as the LOS retrieval. Especially in the context of the temperature retrieval, these spectral lines of CO_2 were selected considering the following requirements:

- Sufficiently strong intensity without being saturated

- Being clearly separable from other spectral signatures

- Different opacity

- Different temperature dependence

While the vertical envelope (i.e. position in altitude) of the retrieved temperature profiles is clearly affected by the previously determined LOS-correction based on the ECMWF temperature profiles, the LOS retrieval does not extract verticaly resolved information from the CO_2-signatures. Furthermore, only a single LOS offset correction parameter or a polynomial correction slowly varying with time was applied for each flight, respectively. In contrast, the temperature retrieval extracts vertically resolved information for every limb scan and thereby significantly improves the temperature information. It is mentioned that for the LOS and temperature retrieval O_3 was also retrieved as additional parameter, as no suited set of CO_2 lines without significant interference with other spectral signatures was available. However, in the chosen microwindows the O_3 signatures are comparably weak and clearly separable from the CO_2 signatures, allowing for a reliable reconstruction of the LOS offset and the temperature profiles. The associated retrieval results for O_3 with comparably low vertical resolution were discarded subsequently, as this gas was retrieved more accurately in dedicated microwindows.

With the knowledge of the LOS correction and the improved temperature profiles, the trace gases were retrieved subsequently according to Figure 3.13. HNO_3 shows only rather weak spectral interference with other species and was therefore retrieved first. Then the retrieval scheme gets branched and the other trace gases were retrieved under consideration of the previously retrieved species. The retrieval scheme can be easily extended to other trace gases: For example HCFC-22 shows a spectral signature around 820 cm^{-1} and can be retrieved under consideration of the previously retrieved profiles of HNO_3 and O_3.

3.2.4 Retrieval result characterisation

For comparisons of the retrieval results with other measurements (i.e. insitu measurements) or model results, knowledge on the retrieval errors and the vertical resolution is mandatory. The MIPAS-STR retrieval results are characterised by the quantifiers (i) estimated combined (random and systematic) 1σ-error, (ii) vertical resolution and (iii) degrees of freedom. The errors in are divided into two groups, the 'primary' errors and 'secondary' errors. The primary errors considered in the estimated combined 1σ-error are:

- spectroscopic line-data uncertainties

- errors in the retrieved temperature profiles

- LOS uncertainties

- radiometric calibration errors (i.e. gain and offset calibration)

- noise error

Errors referred as secondary errors are uncertainties trough (with this list being not exhaustive):

- errors of the pressure profiles interpolated from ECMWF

- uncertainties resulting from horizontal gradients (pressure, temperature, trace gas vmr) along the line of sight

- uncertainties in the trace gas profiles (previously retrieved or climatological) used for forward modelling

- uncertainties of the trace gas distribution above the flight path

- line-mixing of spectral signatures

- uncertainties in FOV characterisation

- ILS model uncertainties

- uncertainties of observer altitude (GPS)

- stray light errors from scan mirror contamination

- errors resulting from the electronic data acquisition chain

While under certain conditions the secondary errors might become important, the consideration of the primary errors resulted predominantly in meaningful errors as confirmed by in-situ comparisions (compare Section 3.2.5).

Uncertainties in the LOS information are discussed in Section 3.1.5 and the combined LOS error is estimated to 1.25 arcmin (1σ). Retrievals with the LOS information biased by this value provided the estimated corresponding LOS-related error for the corresponding target. Uncertainties in the retrieved temperature profiles trough spectral line-data errors for CO_2 and uncertainties in the applied CO_2-profile were estimated by retrievals with a modified CO_2 profile. Thereby, the CO_2 mixing ratios were increased by 5 % of the absolute value at all altitudes, yielding the corresponding error components of the retrieved temperature profiles. Radiometric calibration errors were estimated by retrievals with the spectra scaled by +2 %. The resulting respective individual errors for the retrieved temperature profiles were approximated by the differences between the nominal retrieval result and the modified retrievals (von Clarmann [2003]):

$$\Delta x_j = \begin{pmatrix} \Delta x_{1,j} \\ \Delta x_{2,j} \\ \vdots \\ \Delta x_{n_{\max},j} \end{pmatrix} \approx (\mathbf{K}^T \mathbf{S}_y^{-1} \mathbf{K} + \gamma \mathbf{L}^T \mathbf{L})^{-1} \mathbf{K}^T \mathbf{S}_y^{-1} (f_{\mathrm{error},j} - f_{\mathrm{result}}) \quad (3.7)$$

Here, Δx_j is the profile vector containing the errors associated to the modified quantity j, $\Delta x_{n,j}$ are the errors at the individual altitude levels n, $f_{\mathrm{error},j}$ is the calculated spectrum with the modified quantity and f_{result} the calculated spectrum from the undisturbed retrieval result. The resulting error vectors were then combined with the spectral noise error vector to the estimated combined 1σ-error vector for temperature:

$$\Delta x_{\mathrm{T}} = \sqrt{\Delta x_{\mathrm{prof,spec}}^2 + \Delta x_{\mathrm{los}}^2 + \Delta x_{\mathrm{cal}}^2 + \Delta x_{\mathrm{noise}}^2} \quad (3.8)$$

The individual error contributions $\Delta x_{\text{prof,spec}}$, Δx_{los} and Δx_{cal} resulted from the retrievals with the modified CO_2-profile, LOS and scaling. Δx_{noise} represents the spectral noise error directly obtained from the inversion process.

For the trace gas retrievals, the influence of spectroscopic errors (line-data or cross-sections) were estimated to conservative percentages of the absolute values of the retrieved profiles, assuming that spectroscopic errors are transferred linearly into the retrieval results. For HNO_3, a spectroscopic error of 8 % was considered (compare Wetzel et al. [2002]). The spectroscopic errors for the CFC-11 and CFC-12 cross-sections were set to 10 % considering the results from Moore et al. [2006]. For O_3 and H_2O, spectroscopic errors of 7 % and 10 % were considered, respectively, considering the uncertainties of the line-intensities of the target signatures (compare Flaud et al. [2002]; Flaud et al. [2006] and references therein). For the $ClONO_2$ cross-section an error of 5.5 % was taken into account, considering the results from Wagner and Birk [2003]. Similar to the temperature retrieval, the errors of the retrieved trace gas profiles due to temperature, LOS and radiometric gain errors were approximated by retrievals with modified temperatures (biased by the temperature error), modified LOS and modified scaling. These errors were combined with the spectral noise error to the 1σ-error vector of the considered trace gas retrieval according to:

$$\Delta x_{\text{vmr}} = \sqrt{\Delta x_{\text{spec}}^2 + \Delta x_{\text{temp}}^2 + \Delta x_{\text{los}}^2 + \Delta x_{\text{cal}}^2 + \Delta x_{\text{noise}}^2} \qquad (3.9)$$

with Δx_{temp} the error vector resulting from retrieval with modified temperature (other contributions as in the case of the temperature error).

It is pointed out that the various errors can vary on different timescales and might result in more systematic contributions for different flight sections. While for example the noise error affects single measurements, spectroscopic errors might affect entire flight sections. Therefore, the statistical combination of the different errors by the root of the square sum has to be seen as a simplification, aiming at the combination of the different error components into one single error. Under certain conditions one might also consider the 2σ- or 3σ- values for more intuitive comparisons. The vertical resolutions of the retrieved profiles were analysed considering the averaging kernel matrices $\mathbf{A}$ of the retrievals (Rodgers [2000]):

$$\mathbf{A} = (\mathbf{K}^T \mathbf{S}_y^{-1} \mathbf{K} + \gamma \mathbf{L}^T \mathbf{L})^{-1} \mathbf{K}^T \mathbf{S}_y^{-1} \mathbf{K} \qquad (3.10)$$

Purser and Huang [1993] showed that the reciprocal of the peak of $\mathbf{A}$ can be taken as a measure of its width, which in turn can be seen as a measure for the vertical resolution. Accordingly, in this context the vertical resolution Δa_n at the altitude level n is calculated by

$$\Delta a_n = |\Delta h_n / A_{n,n}| \qquad (3.11)$$

with Δh_n the local vertical retrieval grid spacing and $A_{n,n}$ the corresponding trace element of $\mathbf{A}$.

Finally, the degrees of freedom (DOF) represent the sum of the trace elements of $\mathbf{A}$ and indicate, how many independent pieces of information are derived from the measurement constrained by regularisation (compare Steck [2002]). Hence, the combination of the quantifiers 'combined estimated error', 'vertical resolution' and 'degrees of freedom' provides a comprehensive characterisation of the accuracy, the capability of resolving vertical fine-structures and the a-priori dependence of the retrieval results.

Figure 3.14 shows an example for a retrieved profile from the RECON-CILE flight on March 2nd 2010 associated to a limb scan from a flight altitude of 16.7 km and with a lowest nominal tangent altitude of 9.7 km. The left panel shows the retrieved profile inclusive estimated combined 1σ-error together with the initial guess/a-priori profile. The panel in the middle shows the estimated combined error in percent and the individual individual error contributions. The right panel shows the vertical resolution of the retrieved profile. The retrieved profile shows vertical structures attributed to

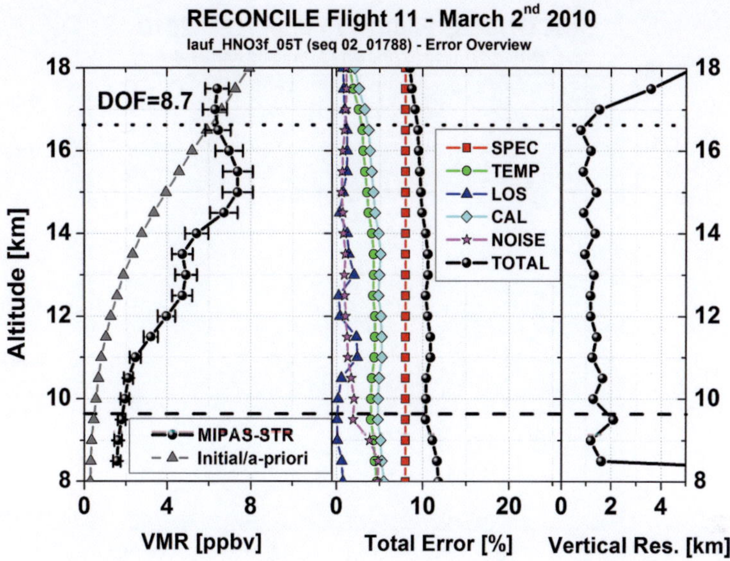

Figure 3.14: Retrieved profile of HNO$_3$, error budget and vertical resolution (from Woiwode et al. [2012]). Left panel: Retrieved profile including estimated combined 1σ-error together with initial guess/a-priori profile (DOF=degrees of freedom). Middle panel: Individual error components and combined error in % (SPEC=spectroscopic error, TEMP=temperature error, LOS=line-of-sight error, CAL=calibration error, NOISE=spectral noise error, TOTAL=combined error). Right panel: Vertical resolution of the retrieved profile. Dotted black line corresponding to flight altitude and dashed black line corresponding to nominal tangent altitude of lowest limb view (center of FOV).

filaments in the vicinity of the polar vortex. The estimated combined errors at the retrieval grid levels are typically between 9 to 12 % and are dominated by the spectroscopic errors. Typical vertical resolutions of 1.0 to 1.5 km are obtained in the vertical range spanned by the nominal tangent points (and slightly above and below). The estimated combined error is valid in context of the vertical resolution, meaning that narrow vertical structures finer than

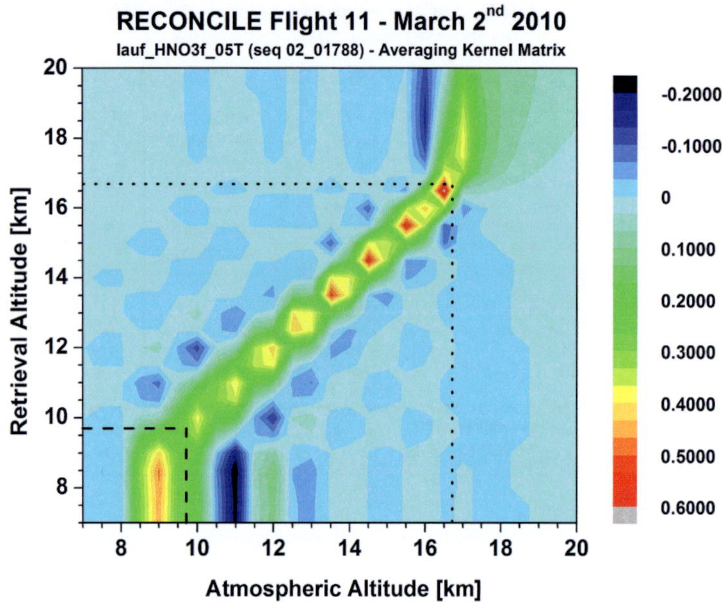

Figure 3.15: Interpolated graphic representation of the averaging kernel matrix associated to the retrieved HNO_3-profile shown in Figure 3.14. Dotted black line corresponding to flight altitude and dashed black line corresponding to nominal tangent altitude of lowest limb view (center of FOV).

the indicated vertical resolution are smoothed out by the retrieval to a certain level. Therefore, local trace gas mixing ratios at certain grid altitudes might be significantly different from the retrieval result if the vertical trace gas distribution shows narrow vertical fine-structures not or only partly resolved by the retrieval. Above 17.5 km practically only column information is obtained and the indicated errors become physically meaningless. Regions below ~8.5 km are outside the FOV for the lowest measurement geometry and therefore no information on the trace gas distribution below is obtained.

Figure 3.15 visualises the averaging kernel matrix of the discussed HNO_3-profile.The averaging kernel matrix indicates the sensitivity of the retrieval at different altitudes. The columns of the matrix indicate the response of the retrieval to a delta function at the indicated atmospheric altitude. The diagonal elements of the matrix are characterised by maxima around the nominal tangent altitudes and visualise the high maximal sensitivities obtained in this vertical range, corresponding to typical vertical resolutions of 1 to 1.5 km. Above flight altitude the diagonal pattern gets strongly broadened and disappears, as mainly column information is obtained of the HNO_3 distribution above. Below $\sim$8.5 km no information on the atmospheric composition is obtained from the measurements, as this region is not covered by sampling.

3.2.5 Geophysical validation of the retrieval results

This subsection describes the validation of the MIPAS-STR measurements with collocated in-situ measurements aboard the Geophysica. Validation is performed for cloud-free conditions at stratospheric altitudes (RECON-CILE Flight 11 on March 2nd 2010) and under the conditions of synoptic-scale PSCs (RECONCILE Flight 5 on January 25th 2010). The in-situ validation under clear sky conditions except for HNO_3 is described in Woiwode et al. [2012] and is summarised in the following. Here, also the validation for HNO_3 is discussed and validation results under PSC conditions are presented. The in-situ instruments[6] the data of which was used for validation and their characteristics are summarised in Table 3.3. Temperature measurements were taken from the Rosemount probe ThermoDynamic Complex (TDC). Ozone measurements were performed by the Fast-

---

[6]TDC and FLASH-A measurements were kindly provided by Sergey Khaykin, Central Aerological Observatory, Dolgoprudny, Moscow region, Russia. FOZAN measurements were kindly provided by Fabrizio Ravegnani, Institute of Atmospheric Sciences and Climate (CNR-ISAC), Bologna, Italy, and by Alexey Ulanovsky, Central Aerological Observatory, Dolgoprudny, Moscow region, Russia. HAGAR measurements were kindly provided by Michael Volk, University of Wuppertal, Germany, and SIOUX measurements by Hans Schlager, German Aerospace Center (DLR), Oberpfaffenhofen, Germany.

Table 3.3: In-situ instruments aboard the Geophysica used for comparisions with MIPAS-STR (from Woiwode et al. [2012], with modifications).

Instrument	Target	Time resolution	Accuracy	Reference
TDC	Temperature	1 s	0.5 K	Shur et al. [2006]
FOZAN	O_3	2 s	$< 10\%$	Ulanovsky et al. [2001]
HAGAR	CFC-11	90 s	0.5–5 ppt (1.4–5.6 %)	Riediger et al. [2000], Werner et al. [2010]
HAGAR	CFC-12	90 s	2–8 ppt (1.2–1.7 %)	Riediger et al. [2000], Werner et al. [2010]
FLASH-A	H_2O (gas)	4 s	10 %	see text
SIOUX	NO_y	1 s	15 %	Voigt et al. [2005]

Response Chemiluminescent Airborne Ozone Analyzer (FOZAN). CFC-11 and CFC-12 measurements were provided by the High Altitude Gas AnalyzeR (HAGAR). Water vapour measurements were obtained by the Fluorescent Lyman-Alpha Stratospheric Hygrometer for Aircraft (FLASH-A), a modified version of the FLASH instrument previously deployed aboard the Geophysica (Sitnikov et al. [2007]). Finally, total gas phase NO_y was measured by the StratospherIc Observation Unit for nitrogen oXides (SIOUX).

For the flight on January 25th 2010 the MIPAS-STR measurements of HNO_3 can be directly compared to SIOUX NO_y, as gas phase NO_y in the Arctic winter lowermost stratosphere under chlorine activated conditions (as supported by the presence of synoptic scale PSCs) is composed almost completely of HNO_3 (i.e. Grooß et al. [2005]; Wiegele et al. [2009] and references therein). Regarding the comparison for March 2nd 2010 it has to be reminded that in the aged vortex air $ClONO_2$ significantly contributes to gas phase NO_y, as seen in the flight sections covering vortex air in the cross-sections in Section 4.2.3.

The MIPAS-STR retrieval results are compared with the in-situ measurements (i) considering MIPAS-STR scans matching approximately the vertical in-situ profile recorded during the ascent phase of the Geophysica and (ii) by comparing the retrieval results at flight altitude with the in-situ measurements along flight track. When comparing the retrieval results with the in-situ measurements, the following circumstances have to be taken into account:

- Match quality: Depending on the flight scenario (flight planning primarily driven by scientific objectives) the spatial coincidence of in-situ and MIPAS-STR measurements is limited.

- Temporal coincidence: For profile comparisons it has to be considered that the MIPAS-STR measurements match the trajectory of the in-situ measurements later in time due to the different measurement geometry. During the time interval in between the atmospheric situation might change.

- Horizontal sampling: In-situ measurements are 1-dimensional measurements resolving horizontal fine structure (i.e. hundreds of meters) constrained to the flight trajectory, whereas the MIPAS-STR measurements integrate significant radiation contributions from several tens (uppermost limb views) to a few hundreds of kilometers (lower limb views) along viewing direction.

- Vertical resolution: While the in-situ measurements resolve vertical fine-structures (i.e. tens of meters), the vertical resolution of the MIPAS-STR measurements is limited (i.e. in the order of kilometers).

Accordingly, the in-situ measurements resolve local vertical and horizontal fine-structures constrained to the flight trajectory, whereas the MIPAS-STR remote measurements integrate properties of more extended atmospheric regions and contain less fine-structure.

In Figure 3.16 the geolocations of the in-situ measurements during the ascent phase of the Geophyisca and the tangent points of the single MIPAS-STR scans selected for comparison are shown for the discussed flights. In the maps also the potential vorticity (PV) from the ERA interim ECMWF analysis is shown at the indicated levels of potential temperature, which approximately correspond to the ceiling altitudes during the discussed flight sections. Airmasses of the polar vortex are characterised by high values of potential vorticity and the vortex edge is indicated according to Nash et al. [1996]. The terms 'potential vorticity' and 'potential temperature' are discussed in Section 4.2. Furthermore, the respective flights are discussed in more detail in Sections 4.2.1 and 4.2.3.

For the in-situ comparison on March 2nd, the PV is shown at the potential temperature level of 450 K, corresponding to an altitude of $\sim$17 km and therefore approximately with the flight altitude in the first half of the flight. As can be seen from the flight trajectory for March 2nd 2010, the in-situ instruments sampled vortex air (i.e. high PV values and inside labeled region) in the first part of the flight and the Geophyisca left the polar vortex at $\sim$78°N. The tangent points of the indicated MIPAS-STR scan were also located inside the vortex, with the instrument however pointing into the PV gradient in the boundary region of the vortex. As the MIPAS-STR measurements integrate radiation contributions from hundreds of kilometers along the viewing direction, the specific viewing situation for this flight has to be considered when interpreting the in-situ comparisons. In the remaining part of the flight the MIPAS-STR measurements covered also filaments from different origin at different altitudes, as discussed in Section 4.2.3. For the retrieval of the considered profile of March 2nd 2010, 9 limb views with tangent altitudes between 9.7 and 16.6 km including upward sampling were available. Thereby, the MIPAS-STR measurements diagonally transected the trajectory of the in-situ measurements with best spatial coincidence at tangent altitudes around 15 km. During the flight on January 25th vortex air was sampled practically during the entire flight (see also Section 4.2.1).

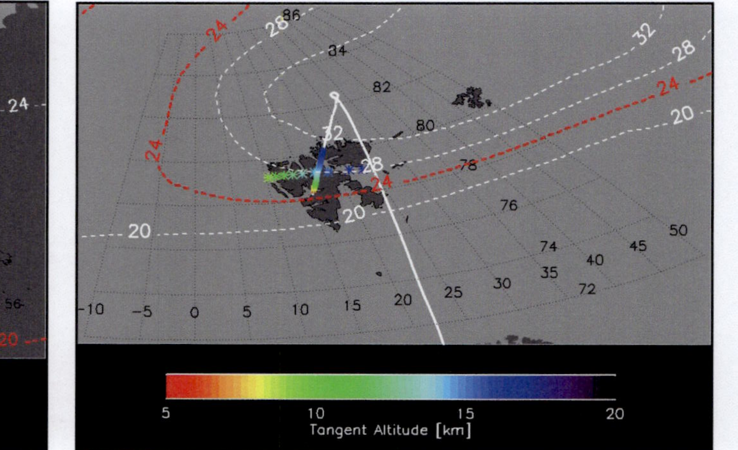
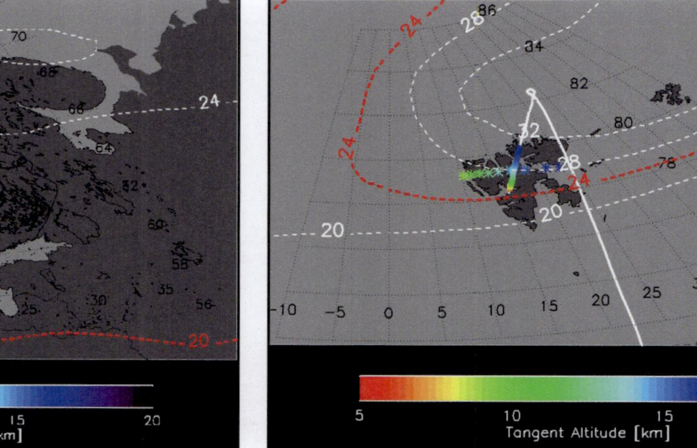

Figure 3.16: Selected MIPAS-STR limb-scans and collocated in-situ measurements for validation. Tangent altitudes of MIPAS-STR measurements indicated asterisks. Flight track of the Geophyisca indicated by white line. Tangent points and flight path associated to in-situ profiles colour-coded with altitude. Left panel: RECONCILE flight 11 on March 2nd 2010. In-situ profile (considered vertical range 8-17 km): 9:41-10:05 UTC. MIPAS-STR scan 02_01788: 11:07 UTC. Right panel: RECONCILE flight 5 on January 25th 2010. In-situ profile (considered vertical range 11-18 km): 5:59-6:30 UTC. MIPAS-STR scan 01_00661: 06:32 UTC. PV-isolines (in PVU) from ECMWF ERA-Interim reanalysis indicated by dashed white lines for the potential temperature level of 450 K at 12:00 UTC for the flight on March 2nd ($\sim$17 km altitude). PV-isolines for the potential temperature level of 430 K at 6:00 UTC for the flight on January 25th ($\sim$18 km altitude). Vortex edge marked by dashed red line according to Nash et al. [1996].

The scan shown in Figure 3.16 was situated well inside the vortex and no steep gradient along viewing direction is identifed in the PV field at a potential temperature of 430 K (corresponding approximately to the flight altitude of 18 km during this scan). For the retrieval of the scan from January 25th 2010, 8 limb views with tangent altiudes between 17.8 and 11.5 km including upward sampling were considered. At this flight the selected MIPAS-STR measurements were shifted parallel by $\sim$10 to 100 km towards south-west compared to the trajectory of the in-situ measurements during ascent.

To overcome the problem that the in-situ measurements may resolve vertical fine structure not resolved by MIPAS-STR, the in-situ results were also smoothed with the corresponding averaging kernels of the retrieved MIPAS-STR profiles. This approach is described by Rodgers [2000] and allows representing the in-situ results smoothed to the vertical resolution of the MIPAS-STR results. The smoothed in-situ profiles x_{s} were calculated according to:

$$x_{\text{s}} = x_{\text{a}} + \mathbf{A}_{\text{lin}} \left(x_{\text{insitu}} - x_{\text{a,lin}} \right) \tag{3.12}$$

with x_{a} the a priori profile of the MIPAS-STR retrieval, $\mathbf{A}_{\text{lin}}$ the averaging kernel of the retrieval linearly interpolated to the in-situ profile altitudes, x_{insitu} the in-situ profile and $x_{\text{a,lin}}$ the a-priori profile of the retrieval linearly interpolated to the in-situ profile altitudes. For the in-situ comparisons along flight track the MIPAS-STR retrieval results at flight altitude are directly compared to the unsmoothed in-situ results.

Figures 3.17 and 3.18 show the results of the in-situ comparisons for cloud-free conditions (March 2nd 2010). For all targets except HNO_3 both the profile comparison and the comparison along flight track are shown. From SIOUX too few in-situ datapoints were available from the ascent phase. The profile comparisons (left side) indicate reasonable agreement between the MIPAS-STR and in-situ measurements for T, O_3 and H_2O within the combined errors of MIPAS-STR and the in-situ measurements.

The uncertainties of the in-situ measurements are not shown in the plots but are given in Table 3.3.

For CFC-11 and CFC-12 the MIPAS-STR and HAGAR profiles are also mostly in agreement, while these gases seem to be notably overestimated by MIPAS-STR at altitudes above 14 to 15 km. This fact is explained by the circumstance, that the corresponding measurements of MIPAS-STR during this flight section integrated also contributions from vortex-edge region characterised by higher mixing ratios of the CFCs along their extended viewing geometries. As the CFCs show largest contrasts in their mixing ratios between vortex and extra-vortex air[7], the MIPAS-STR measurements of the CFCs are significantly affected by the horizontal gradient and contrast in the corresponding mixing ratios at the vortex edge. For example O_3 shows much weaker contrasts in mixing ratios between inside and outside the vortex and therefore the MIPAS-STR results match the in-situ datapoints much better. Accordingly, the differences between the MIPAS-STR and in-situ profiles of the CFCs are attributed to strong horizontal inhomogeneities in the mixing ratios of the CFCs along the MIPAS-STR viewing geometries and are not interpreted as systematic offset of the MIPAS-STR results. The comparisons along flight track for the flight on March 2nd 2010 yield similar results: The temperature, O_3 and H_2O measurements of MIPAS-STR and the in-situ instruments are in excellent agreement during the entire flight. These results indicate the absence of systematic errors during the entire flight and underline the constant quality of the MIPAS-STR measurements. For the CFCs, also the major structures and several fine structures along the flight track found in the in-situ results (i.e. vortex edge between 11:00 and 11:20, indicated by steep increase in the mixing ratios of the CFCs) are reasonably reproduced by MIPAS-STR. Again, during the first flight part, with the in-situ measurements exclusively sampling vortex air, MIPAS-STR also integrated radiation contributions from

---

[7]The CFC-11 mixing ratios approach lowest values at 17 km altitude inside the vortex, compare Figure 3.18.

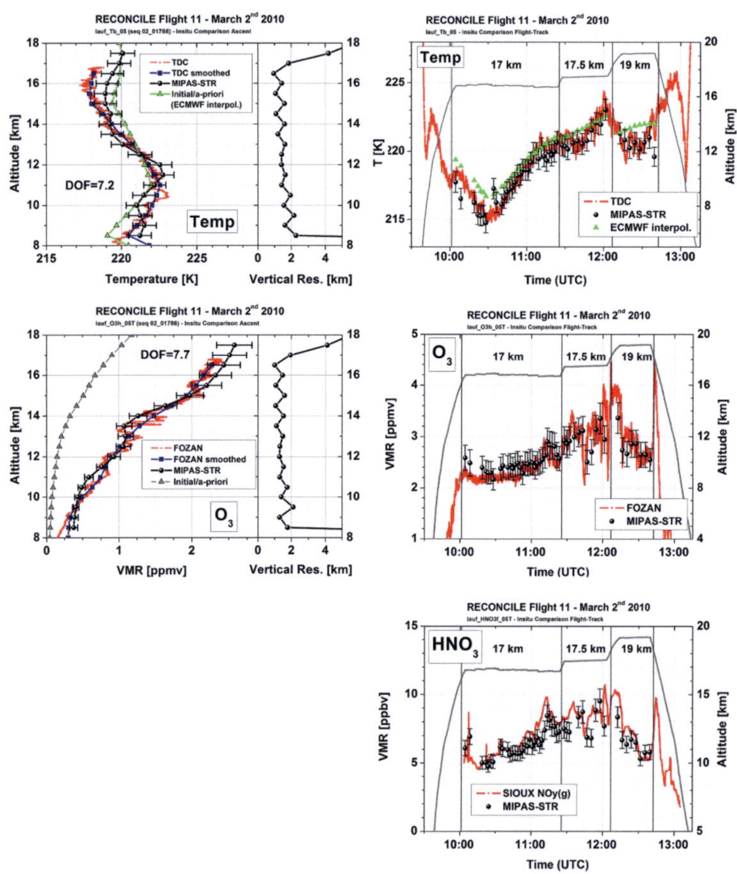

Figure 3.17: In-situ comparisons for temperature, O_3 and HNO_3 under cloud-free conditions (March 2nd 2010). Left side: MIPAS-STR single retrieved profiles, a-priori/initial guess profiles and vertical resolution of retrieved profile. The in-situ profiles were also smoothed with the averaging kernels of the corresponding MIPAS-STR profiles. Right side: MIPAS-STR retrieval result at ~flight altitude and in-situ measurements at flight altitude. MIPAS-STR results for HNO_3 compared with total gas-phase NO_y from SIOUX. (all plots except for HNO_3 from Woiwode et al. [2012]).

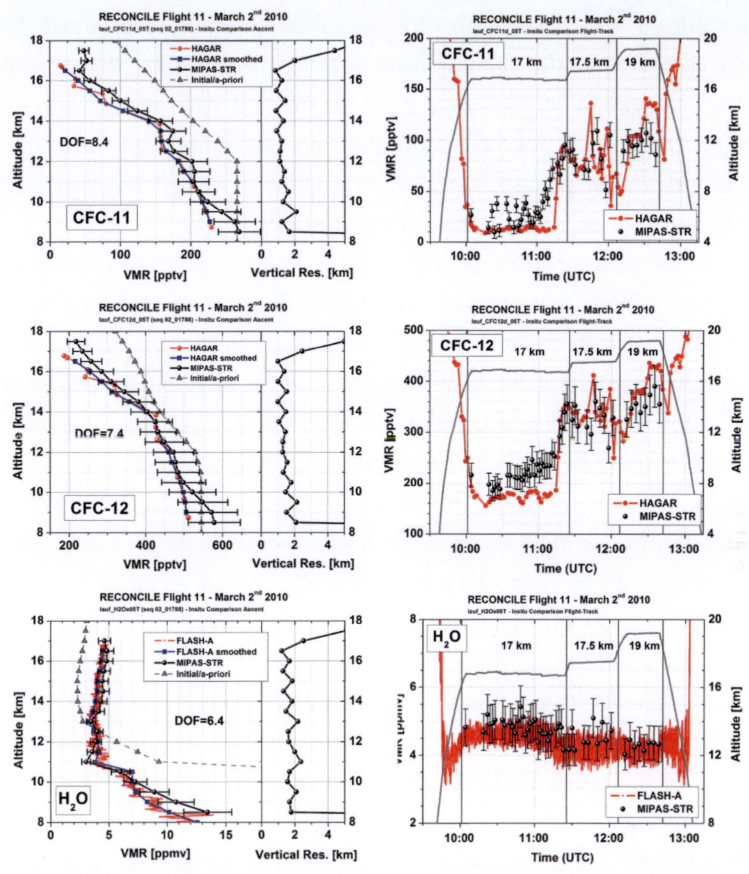

Figure 3.18: As Figure 3.17 but for CFC-11, CFC-12 and H_2O (from Woiwode et al. [2012]).

the vortex edge region characterised by higher mixing ratios of the CFCs. The same systematics with different prefix also applies for the time interval around 12:30 UTC: While the in-situ instruments probed extra-vortex air, the MIPAS-STR measurements also integrated radiation from vortex filaments along the viewing direction, as also discussed in Section 4.2.3.

Therefore, lower mixing ratios for the CFCs are obtained from MIPAS-STR compared to the in-situ measurements. The absence of systematic offsets in the MIPAS-STR results for the CFCs is also supported by the finding that the MIPAS-STR retrieval results show both higher and lower mixing ratios compared to HAGAR, depending on whether the instrument looks from vortex air into extra-vortex air or vice versa.

The in-situ comparison of MIPAS-STR HNO_3 and SIOUX gas-phase NO_y along flight track shows a high degree of agreement between the involved instruments and fine structures along flight track are reproduced (i.e. maximum around 11:15 UTC). However, it has to be reminded that inside the vortex air the total NO_y measured by SIOUX also includes significant contributions of $ClONO_2$ due to previous deactivation of active chlorine species in the aged vortex. The MIPAS-STR retrieval results for $ClONO_2$ associated to this flight discussed in Section 4.1.3 (compare Figure 4.10) show that in the flight section until 11:00 $ClONO_2$ mixing ratios as high as 1.5 ppbv are present. Considering the retrieved HNO_3 mixing ratios along flight track and the associated $ClONO_2$ mixing ratios, the sum of these gases seems to be overestimated by MIPAS-STR compared to SIOUX. This finding can be explained also by effects from horizontal inhomogeneities: The horizontally extended MIPAS-STR observations probably covered significant portions of less denitrified air (or more renitrified air, respectively), whereas the SIOUX measurements were recorded in more denitrified air. However, the comparison during the remaining part of the flight characterised by higher retrieved HNO_3 mixing ratios and less $ClONO_2$ show reasonable agremeement between MIPAS-STR and SIOUX.

As this section also summarises the typical characteristics of the individual retrievals (combined error, vertical resolution and DOF), the retrieved $ClONO_2$ profile for the discussed limb scan from March 2nd is shown in in Figure 3.19. The profile shows high $ClONO_2$ mixing ratios of about 1200 pptv at 16 km altitude and above, indicating chlorine-deactivated air in the late polar vortex. Virtual negative retrieved mixing ratios at lowest altitudes

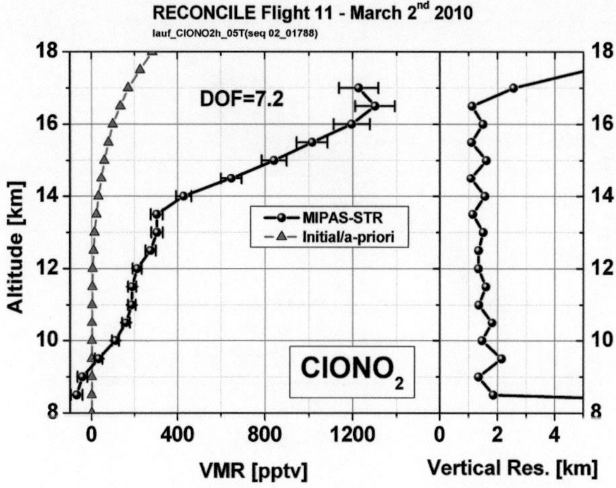

Figure 3.19: Left side: Retrieved vertical profile of ClONO$_2$ from scan 02_01788 on March 2nd 2010 with estimated combined error, DOF and initial guess/a priori profile. Right side: Vertical resolution of the retrieval result (from Woiwode et al. [2012]).

result from systematic addition of different error components and smoothing effects from the retrieval, which is characterised by a low sensitivity on this specific gas at these altitudes.

To overcome the limitations of the comparisons due to different sampling characteristics in Figure 3.20 the correlations between CFC-11 and CFC-12 derived from the MIPAS-STR and HAGAR in-situ measurements are shown. As these gases have long atmospheric live times (about 50 years for CFC-11 and 100 years for CFC-12) and the sources of these gases vary slowly (Montzka et al. [2011]) their mixing rations can be seen as approximately constant in vortex and extra-vortex air and the therefore the in-situ comparisons should be less affected from situations when MIPAS-STR samples different airmasses within single observations. This approach makes the in-situ comparisons less sensitive on contributions from vortex-

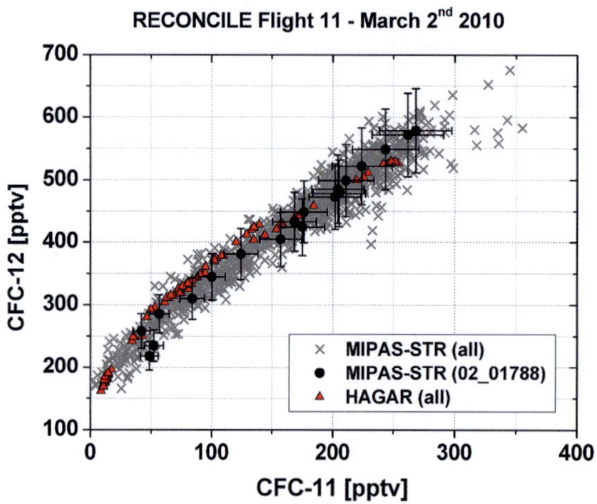

Figure 3.20: Correlations of CFC-11 and CFC-12 derived from MIPAS-STR (all scans from the flight and individual scan 02_01788 associated to in-situ profile comparisons) and HAGAR (all data points) for the flight on March 2nd 2010 (from Woiwode et al. [2012]).

edge/extra-vortex air integrated by the MIPAS-STR measurements. The correlations of the CFCs derived from MIPAS-STR and HAGAR shown in Figure 3.20 indicate enhanced agreement between the two instruments. HAGAR shows slightly higher mixing ratios of CFC-12 versus CFC-11 for CFC-11 mixing ratios between 30 and 150 pptv. However, these datapoints are also situated within the scatter of the MIPAS-STR datapoints from the entire flight.

Figure 3.21 shows in-situ comparisons for the flight on January 25th 2010 under the conditions of synoptic scale PSCs. At altitudes above 17 to 18 km PSCs were detected by in-situ and remote sensing measurements (compare Sections 4.2.1 and 4.2.3). Between ascent and descent the entire flight was carried out inside PSCs. The comparison of the MIPAS-STR temperature profile (left side) with the in-situ measurements of TDC during ascent

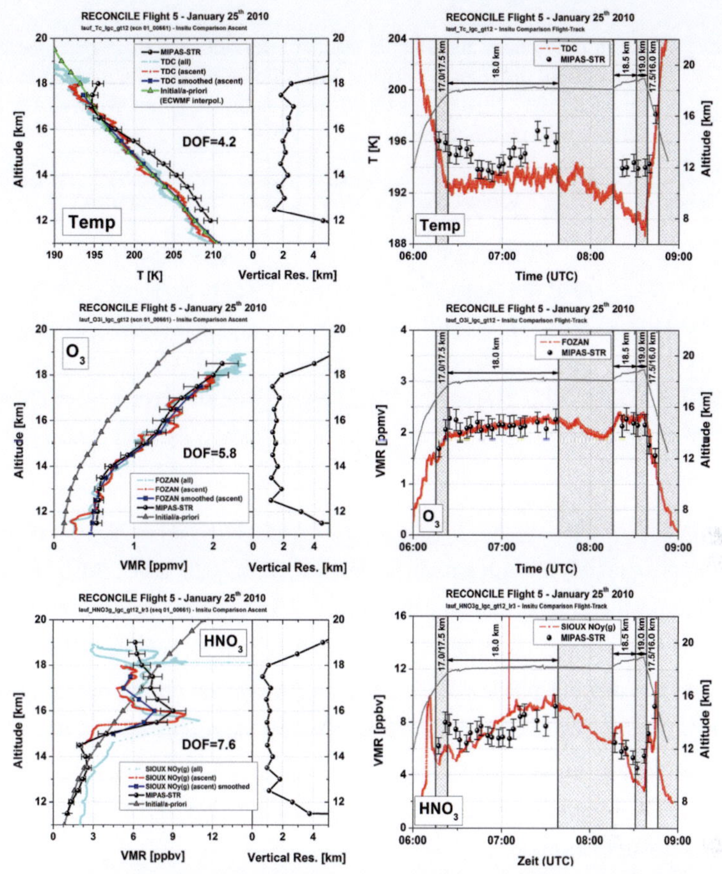

Figure 3.21: In-situ comparisons under PSC conditions (January 25th 2010). Legend according to Figure 3.17. Modification on left side: in-situ measurements also indicated for entire flight in (cyan), including also descent profile (see text).

shows best agreement between 15.5 and 17 km. At altitudes above and below the MIPAS-STR measurements indicate temperatures enhanced by 1 to 3 K. Also the comparison along flight track (left side) mostly shows significantly higher temperatures for MIPAS-STR (up to 4.5 K), while the results

Table 3.4: Mean differences and sample standard deviations (1σ, in parentheses) of the mean differences between MIPAS-STR and in-situ measurements on March 2nd 2010 under cloud-free conditions (from Woiwode et al. [2012], with modifications).

Target	Vertical profile		Flight track	
T	0.21 (0.73) K		−0.19 (0.64) K	
O_3	−0.01 (0.12) ppmv	−3 (14) %	0.05 (0.26) ppmv	2 (9) %
CFC-11	17 (9) pptv	14 (13) %	8 (24) pptv	21 (40) %
CFC-12	21(19) pptv	5 (5) %	14 (42) pptv	8 (16) %
H_2O	1.16 (1.91) ppmv	8 (14) %	0.29 (0.36) ppmv	6 (7) %
HNO_3	-	-	-0.60 (0.81) ppbv	-9 (11) %

are approximately in agreement between 6:40 to 7:20 UTC. Also shown for the profile comparison are the temperature measurements from TDC for the entire flight, including also the vertical temperature profile associated to the descent phase. The in-situ measurements during descent also matched aimasses similar to those sampled by the discussed MIPAS-STR scan (compare Figure 3.16). These measurements indicate that at ceiling altitude TDC sampled also warmer regions fitting better to the MIPAS-STR profile at flight altitude.

The in-situ comparison for temperature can be interpreted as follows: As TDC sampled also regions with higher temperatures along flight track, approximately matching those in the upper part of the MIPAS-STR profile, the discrepancy between MIPAS-STR and the in-situ measurements might be explained by horizontal inhomogeneity of the atmospheric temperature field. The same applies for the finding that the MIPAS-STR temperatures along flight track are mostly higher than the in-situ temperatures, while some of the data points of MIPAS-STR match the in-situ measurement. As can be seen in Section 4.2.1, Figure 4.1, MIPAS-STR sampled

Table 3.5: Mean differences and sample standard deviations (1σ, in parentheses) of the mean differences between MIPAS-STR and in-situ measurements on January 25th 2010 under PSC conditions.

Target	Vertical profile		Flight track	
T	1.59 (1.01) K		2.31 (1.08) K	
O_3	−0.03 (0.05) ppmv	−2 (6) %	0.00 (0.14) ppmv	0 (7) %
HNO_3	1.18 (0.89) ppmv	14 (14) %	0.20 (1.30) ppbv	4 (20) %

in the time-interval 6:15 to 6:30 UTC and 6:30 to 7:20 UTC similar airmasses form different (perpendicular) directions, resulting in the different temperatures along flight track shown in figure 3.21. Accordingly, during this flight strong temperature gradients along the viewing direction play an important role and can explain the different temperatures observed for approximately the same airmasses sampled from different directions. It has to be pointed out that a temperature offset in the MIPAS-STR temperature retrieval results cannot be excluded. However, the excellent agreement of the in-situ comparison for ozone and also the reasonable agreement for MIPAS-STR HNO_3 versus SIOUX NO_y (see below) reduces the likelihood of such an offset, as the corresponding retrievals consider the previously retrieved temperatures.

The in-situ comparison for O_3 shows excellent agreement for the vertical profile and the comparison along flight track. This finding is again explained by the fact that this gas shows only relative weak horizontal gradients and contrasts in mixing ratios affecting the MIPAS-STR observations. The comparison of retrieved MIPAS-STR HNO_3 with in-situ gas phase NO_y also shows a high degree of agreement. Discrepancies are here also explained by a rather inhomogeneus horizontal distribution of gas phase HNO_3, as airmasses inside and directly below PSC clouds were probed, with significant amounts of HNO_3 in the condensed phase. For the profile comparison, also the in-situ NO_y measurements for the entire flight are shown and support a rather inhomogeneous distribution of gas phase NO_y in horizontal directions around flight altitude. However, the MIPAS-STR measurements again cover a similar range as the in-situ measurements, supporting that no obvious systematic offset in the MIPAS-STR measurements is present. Furthermore the NO_y profile sampled by SIOUX with a characteristic maximum around 16 km altitude is reproduced in the MIPAS-STR HNO_3 profile to a high degree. Therefore, the shown comparison confirms the capability of MIPAS-STR measurements of resolving narrow vertical structures.

Consequently, also the MIPAS-STR retrieval results for January 25th under PSC conditions can be conciliated with the collocated in-situ measurements taking into account an inhomogeneous atmospheric scenery. In-situ comparisons for CFC-11 and CFC-12 were not carried out for this flight as too few correlative in-situ data was available. The H_2O retrieval for this flight was discarded, as the measured spectral signatures of this gas are potentially affected by stray light inside the PSCs[8].

Tables 3.4 and 3.5 summarise the results of the in-situ comparisons and show that the MIPAS-STR results are mostly in reasonable agreement with the corresponding in-situ measurements within the errors of MIPAS-STR and the involved in-situ instruments (compare Table 3.3) except for the discussed particular cases.

The retrieved profiles for both cloud-free and cloud affected conditions show typically combined errors between 10 to 15 % and typical high vertical resolutions between 1.0 and 2.5 km. For the representative limb scan under cloud-free condtions (Figures 3.17 and 3.18) with 9 tangent altitudes between 9.7 and 16.6 km (including upward sampling) between 6.4 to 8.7 DOF are obtained for the different retrieval targets. This corresponds to

---

[8] H_2O shows a rather strong increase at tropospheric altitudes by two orders of magnitude; accordingly significant emissions from H_2O might have been scattered into the MIPAS-STR measurements by PSC particles.

$\sim$0.7 to 1.0 DOF per tangent altitude, while also information is contributed by the upward viewing measurements. Similar values for DOF per tangent altitude are obtained for the scan under PSC conditions (Figure 3.21) for O_3 and HNO_3, while the necessity of a stronger regularisation resulted in a slightly lower value of $\sim$0.5 DOF per tangent altitude for temperature.

3.2.6 Summary: Retrieval strategy and characterisation of level-2 product

A consistent retrieval strategy for cloud- and aerosol-affected MIPAS-STR measurements from the UTLS region has been elaborated. High vertical resolutions in the order of 1 kilometer are feasible, allowing detailed and quantitative studies on the chemical and dynamical situation in this vertical range and providing the basis for the studies on dentrification and chlorine deactivation presented in the following. Accurate retrievals are shown to be possible from stratospheric observations with cloud index values as low as 1.6 (see also Section 4.2.1). The applied retrieval strategy is based on a sequential retrieval, avoiding the simultaneous reconstruction of many different parameters. Wave number-independent continuum extinction was inverted for each retrieval microwindow to compensate for the effects of clouds, aerosol and broad signatures from non-specified trace gas emissions. A set of microwindows has been exploited and verified, allowing reliable retrievals of the target parameters under the conditions of the Arctic winter UTLS region. The retrieval results are validated with collocated independent in-situ measurements under cloud-free conditions and conditions affected by PSC clouds. The MIPAS-STR retrieval results are characterised as follows:

- Retrieval error: The estimated combined (random and systematic) 1σ-errors at the retrieval grid levels including spectroscopic errors are typically between 10 to 15 % for trace gas mixing ratios and < 1 K for temperature.

- Vertical resolution: Typical vertical resolutions between 1 to 2.5 km are obtained between flight altitude and the lowest measurement geometry depending on the target parameter and the atmospheric conditions.

- For the flight on March 2nd typical DOF values between 6.4 and 8.7 are found for the discussed scan with a vertically resolved region between 8.5 and 17.0 km. Typical DOF values per tangent altitude (including also contributions from upward sampling) between 0.5 and 1.0 are obtained for the discussed scans under cloud-free and PSC-affected conditions.

- Validation: The retrieval results show reasonable agreement with collocated in-situ measurements of temperature, O_3, CFC-11, CFC-12, H_2O and NO_y, mostly within the errors of the involved instruments. Discrepancies found for temperature and HNO_3 (versus in-situ NO_y) for the flight on January 25th 2010 and to a lower extent also for HNO_3, CFC-11 and CFC-12 for the flight on March 2nd 2010 are explained by horizontal inhomogeneities, as remote sensing measurements are compared with localised in-situ measurements. A potential warm bias in the retrieved temperatures for the flight on January 25th cannot be excluded. However, the observed discrepancies might also be explained the variability of stratospheric temperatures during this flight, as indicated by the MIPAS-STR measurements from the whole flight (compare Section 4.2.1).

- The overall constant degree of agreement indicated by the in-situ comparsions for entire flights confirms that the calibrated measurements and the retrieval results are characterised adequately.

4 Scientific results

This chapter discusses the scientific evaluation of the MIPAS-STR measurements from the RECONCILE field campaign in context of Arctic ozone depletion chemistry, PSC microphysics and atmospheric dynamics.

In the first part of this chapter, three flights covering the period between end of January until begin of March 2010 are analysed. The discussed flights cover different phases of the Arctic stratospheric winter 2009/10, including measurements inside polar stratospheric clouds and chlorine activated air in January and aged deactivated vortex air in the beginning of March. The retrieved profiles from MIPAS-STR are first combined to 2-dimensional vertical cross-sections of temperature, trace gas mixing ratios and cloud coverage along flight track and the observed patterns are discussed in Sections 4.2.1 to 4.2.3. The vertical cross-sections give insight into the chemical and dynamical situation at stratospheric altitudes during the discussed flights. Thereby, the measurements demonstrate the capability of this measurement technique of resolving 2-dimensional mesoscale atmospheric structures. The retrieval results of HNO_3 and $ClONO_2$ are brought into a quantitative context in Section 4.2.4. Correlations with the long-lived tracer CFC-12 are used to study denitrification and chlorine deactivation.

The second half of this chapter discusses a detailed study on denitrification by potential NAT particles in the Arctic winter 2009/10. Large-dimension PSC particles potentially consisting of NAT and capable of denitrification were observed by in-situ measurements during RECONCILE (von Hobe et al. [2012]). The observed particles sizes can hardly be reconciled with current theory of nucleation and growth of compact spherical

NAT particles. This study explores the hypothesis that the observed particles were significantly aspheric and/or had low particle mass densities, allowing the growth of particles with large maximum dimensions in relative short time. Such particles would have slower settling velocities due to a larger surface and therefore increased friction in the stratospheric air. Accordingly, the vertical and horizontal patterns of HNO_3 redistribution are modified for NAT particles settling more slowly. Different scenarios of the chemical transport model CLaMS are analysed, considering denitrification by NAT particles with reduced settling velocities. Thereby, the reduced settling velocities simulate the sedimentation of aspheric or 'flake-like' (i.e. low particle mass density) NAT particles. The resulting modeled patterns of HNO_3 redistribution through denitrification are compared with the MIPAS-STR measurements, and conclusions are drawn on the properties of potential stratospheric NAT particles.

Overall, the combination of the MIPAS-STR measurements, the in-situ measurements and the model simulations discussed in the following subsections gives insight into the properties of the Arctic polar vortex in the winter 2009/10 and into details of the processes of denitrification and chlorine deactivation.

4.1 The Arctic winter 2009/10 and the RECONCILE field campaign

An overview of the meteorological conditions during the Arctic stratospheric winter 2009/10 is given by Dörnbrack et al. [2012]: A quick and early cooling of the early polar vortex was prevented until begin of December due to planetary wave activity. In early December, a first vortex split occurred, which was followed by the formation of a strong and cold vortex from the middle of December 2009 until the end of January 2010. During this mid-winter period the vortex was exceptionally cold and allowed the existence of synoptic scale PSCs until the end of January. The cold phase was ended by a sudden stratospheric warming at the end of January with the vortex being markedly displaced from the pole. The vortex split up into two lobes, of which one lobe remained compact until mid of March. While the mid-winter phase was exceptionally cold with temperatures below the climatological mean, the overall Arctic winter 2009/10 was one of the warmest Arctic winters in the last 21 years.

RECONCILE was an integrated European research project, aiming on improving the understanding of Arctic stratospheric ozone loss and climate interactions in order to improve simulations of the evolution of the stratospheric ozone layer in the future (von Hobe et al. [2012]). The key objectives of RECONCILE were:

- a quantitative understanding of polar stratospheric ozone depletion chemistry

- improved understanding of PSC microphysics and heterogeneous chlorine activation

- quantification of fluxes through the polar vortex edge and influence on ozone in the midlatitudes

- improved parameterisations for Climate Chemistry Models (CCMs) for enhanced predictability of the evolution of the ozone layer in future and climate interactions

To achieve these goals RECONCILE involved laboratory measurements, field campaigns and atmospheric chemistry modelling.

The RECONCILE field campaign was based at Kiruna airport, Sweden (67°49'N/ 20°20'E) and was carried out in two parts: The first part of the campaign included research flights of the Geophysica between January 17th and February 2nd 2010, allowing sampling of PSC clouds and chlorine-activated air during the unusually cold period of this winter. The second part included research flights of the Geophysica between February 27th and

March 10th 2010 and allowed probing of the remaining aged vortex lobe and vortex filaments. A relay flight was carried out with an intermediate landing in Spitsbergen, Norwegen, probing the aged vortex lobe moving from Canada towards Siberia.

4.2 MIPAS-STR measurements in Arctic winter 2009/10: Denitrification and chlorine deactivation

In the following, MIPAS-STR retrieval results are discussed for three selected RECONCILE flights:

- **January 25th 2010**: This flight was carried out at the end of the exceptional cold phase of the 2009/10 Arctic winter under the conditions of synoptic scale PSCs.

- **January 30th 2010**: Also this flight was carried out at the end of the exceptional cold phase of this Arctic winter at the beginning of the sudden stratospheric warming.

- **March 2nd 2010**: This flight probed the remaining compact lobe of the aged vortex situated above Spitsbergen after the sudden stratospheric warming and vortex split.

The combination of these flights allows to study the processes dentrification and chlorine deactivation in the late cold phase under the conditions of PSCs and cloud-free conditions and furthermore in the aged vortex after the vortex split.

For each flight, the flight path of the Geophysica and the horizontal distribution of the MIPAS-STR tangent points are shown in the following. The the situation of the polar vortex is indicated by showing the distributions of the potential vorticity provided by the ECMWF ERA-Interim reanalysis for the specific flights at levels of constant potential temperature corresponding approximately to the flight altitude[1].

---

[1] PV-fields: Courtesy Jens-Uwe Grooß, Research Centre Jülich GmbH, Germany)

The potential vorticity (i.e. Holton [2004]) is a measure for the rotation of the flow field in meteorological analyses and the corresponding units are the potential vortiticy units (PVU, with 1 PVU = 10^{-6} K m^2 kg^{-1} s^{-1}). The potential vorticity is calculated according to

$$PV = \rho^{-1} \left(\nabla \times v + 2w \right) \nabla \Theta \tag{4.1}$$

with ρ being the air density, v the three-dimensional velocity field, w the angular velocity of the earth and Θ the potential temperature. Lines of constant PV at surfaces with constant potential temperature represent trajectories of adiabatic motion in the atmosphere.

The potential temperature corresponds to the theoretical temperature of an air parcel that is brought adiabatically to a standard pressure of 1000 mbar (i.e. Holton [2004] and references therein) and is calculated according to

$$\Theta = T \left(\frac{p_0}{p} \right)^{\left(\frac{R}{c_p} \right)} \tag{4.2}$$

with p being the atmospheric pressure, p_0 the standard pressure of 1000 mbar, R the gas constant (R=8.31451 J K^{-1} mol^{-1}) and c_p the heat capacity at constant pressure (c_p=29.1 J K^{-1} mol^{-1}). The potential temperature is an important quantity in atmospheric dynamics, as it is not affected by lifting and sinking of airmasses, for example in the presence of mountains or by large-scale turbulence.

The airmasses of the polar vortex are characterised by high values of the PV. Nash et al. [1996] showed that the edge of the polar vortex can be identified considering the PV distribution (i.e. the maximum gradient) and the location of the maximum wind jet. In the plots showing the flight paths of the Geophyisca in the following, the vortex edge is indicated according to the Nash-criterion.

For each flight the vertical cross-sections of the cloud-index are shown in the following to indicate the vertical and horizontal cloud coverage. For the flight on January 25th also the retrieved vertical cross-section of continuum is indicated, showing fine-structures at altitudes affected by PSCs. Then, for each flight the retrieved vertical cross-sections of temperature and trace gas mixing ratios are discussed, showing characteristic patterns associated to the polar vortex and extra-vortex air. Finally, the retrieval results for HNO_3 and $ClONO_2$ are correlated with the retrieved CFC-12, and dentrification and chlorine deactivation are discussed for the period between the end of January and begin of March 2010.

4.2.1 Flight on January 25th 2010: Cold vortex under synoptic scale PSC conditions

The flight on January 25th was carried out at the end of the PSC phase of the Arctic winter 2009/10. The flight conditions were characterised by a compact and cold polar vortex (Dörnbrack et al. [2012]). Synoptic scale PSC clouds consisting of STS and NAT-containing mixtures were detected by the CALIPSO space-borne lidar during the late January period (Pitts et al. [2011]).

The Geophysica flight on January 25th 2010 took place in the morning from 5:50 to 9:19 UTC with takeoff and landing in Kiruna, Sweden. The flight track of the Geophysica and the MIPAS-STR tangent points are indicated in Figure 4.1 together with the meteorological context: The potential vorticity (PV) isolines are characterised by low values, indicating that the whole flight and sampling was carried in vortex air. Except for scan B_1 the tangent points of all scans were situated well within the polar vortex according to the criterion of Nash et al. [1996], as indicated by the PV distribution at about flight altitude. To avoid blending of the detectors by the

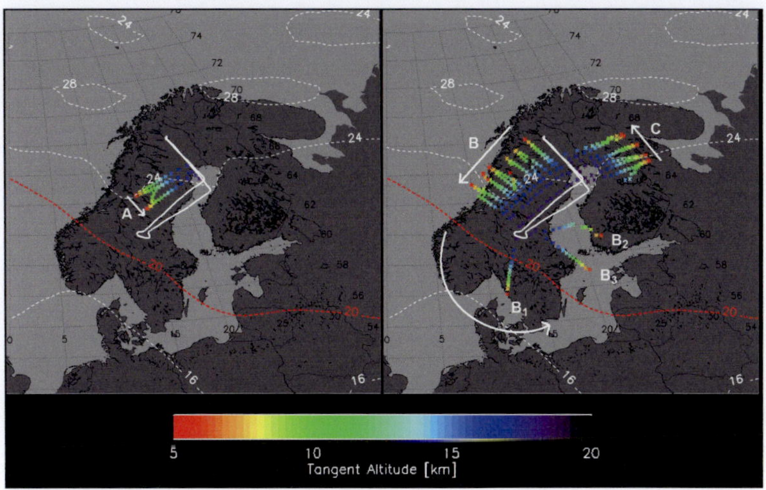

Figure 4.1: Flight track of the Geophysica during RECONCILE flight 5 on January
25th 2010. Tangent points of MIPAS-STR observations colour-coded
with altitude. PV-isolines (in PVU) at the potential temperature level of
430 K ($\sim$17 km) at 6:00 UTC from ECMWF ERA-Interim reanalysis
indicated by dashed white lines. Vortex edge marked by dashed red line
according to Nash et al. [1996].

rising sun (in $\sim$south-east direction), in the time interval between 7:40 to
8:15 UTC no limb-scanning was performed.

The vertical distribution of the MIPAS-STR tangent points and the flight
altitude of the Geophysica are shown in Figure 4.2 together with the interpolated cloud index (compare Section 3.2.1). After the ascent phase the
flight was carried out at a constant altitude of $\sim$18 km, with the Geophyisca
climbing in the last part of the flight (from 8:15) to a ceiling altitude of 19
km.

The interpolated cloud index shows low values between 1 and 4 for the
entire flight and indicates the influence of PSC particles on all measurements. In particular around flight altitude low cloud index values around
2 are found, clearly pointing at the presence of PSC particles. In the last

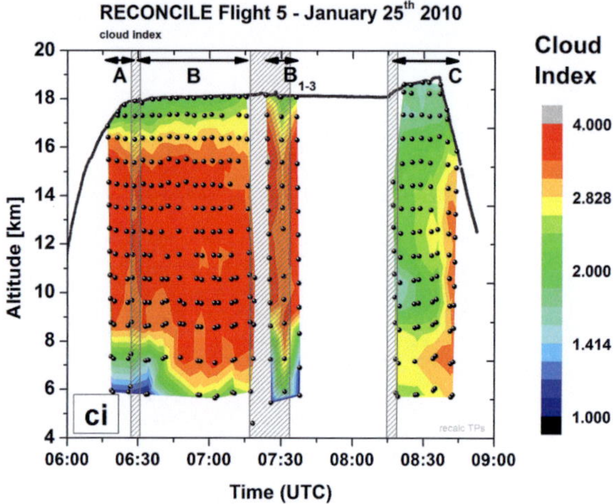

Figure 4.2: Flight altitude, vertical distribution of MIPAS-STR tangent points and cloud index for the flight on January 25th 2010.

flight section, cloud index values as low as 1.6 are observed at flight altitude and below. At intermediate altitudes (<17 km) cloud index values close to 4 are found in the first half of the flight indicating cloud-free conditions. However, these measurements with relatively low cloud index values are still affected by the PSC particles present at higher altitudes.

At lower altitudes <9 km again lowest cloud index values are observed in sections A and B as a consequence of tropospheric clouds. In the last flight section C, between 8:15 and 8:40 UTC, a separate cloud index minimum is found around 10 km, probably indicating a partially transparent cirrus cloud. Higher cloud index values below indicate a relatively cloud-free troposphere. It has to be mentioned that in this graph especially the contrasts in the cloud index give information on the presence of clouds, while the absolute cloud index values are biased to some extent by the optical thickness of the PSC layers around flight altitude. Accordingly, the

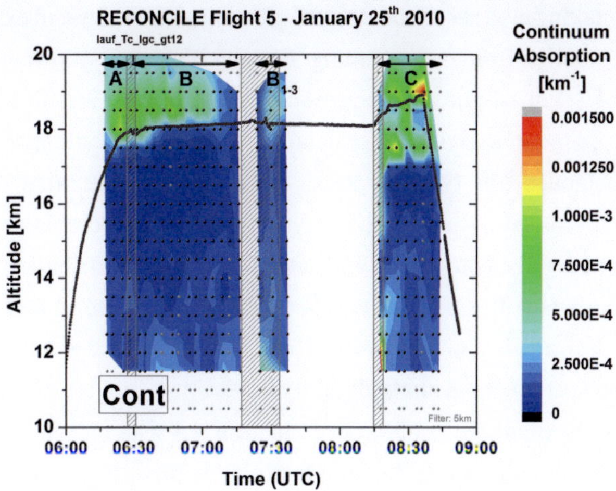

Figure 4.3: Flight altitude and retrieved vertical cross-section of continuum extinction, qualitatively indicating vertical PSC coverage for the flight on January 25th 2010. Grey hatched areas marking turns performed by the Geophysica with less reliable pointing information potentially affecting the MIPAS-STR retrievals. Retrieved gridpoints indicated by dots are filtered for vertical resolution better than 5 km (black dots, typical vertical resolution ∼1 km) for interpolation.

cloud index values around 2 at altitudes around 14 km in section C probably hint on cloud-free conditions, whereas values of ∼1.5 around 10 km are interpreted to result from a cirrus cloud. It is mentioned that in case of conservative cloud-filtering (compare Spang et al. [2004]) practically all spectra of this flight were discarded. However, applying the retrieval strategy discussed in Section 3.2.3 allowed to make comprehensive use of the measurements during this flight.

Figure 4.3 shows the vertical distribution of continuum extinction retrieved from the spectra. The plot shows continuum extinction retrieved along with the temperature retrieval in the 810.1–813.1 cm^{-1} microwindow. The continuum extinction is a retrieved quantity and is reconstructed on the finer grid, making also use of the upward viewing geometries and extracting further information from oversampling. Locally high vertical resolutions in the order of 1 km (slightly better around flight altitude) are obtained for continuum extinction, giving more detailed insight into vertical structures of the PSC clouds (the typical vertical resolutions for the other retrieval targets are discussed in Section 3.5.2). While the retrieved continuum extinction alone is only a qualitative indicator showing the distribution of continuum absorbers with sufficient number density, it is a useful tool to gain qualitative information on the vertical and horizontal extent of optically partially transparent PSCs along flight track.

Significantly enhanced continuum extinction pointing on PSCs is found around flight altitude during the entire flight, with stronger maxima between 6:15 and 7:10 UTC (sections A and B) at flight altitude and above. Between 8:15 to 8:45 UTC (section C) maxima are found below and slightly above flight altitude. The pattern in the latter region indicates overlaying PSC layers with relatively low continuum extinction in between, with these layers appearing vertically connected at locations around 8:30 UTC. A sharp contrast between strongly enhanced continuum extinction, indicating PSC clouds, and cloud-free regions with low continuum extinction is observed. Thereby the retrieved continuum extinction allows more accurate assignment of PSC-affected layers than the cloud-index plot in Figure 4.2, which serves as a first-order indicator for particle-affected layers. Further regions with slightly enhanced continuum extinction are found at lower altitudes and might hint on cirrus-clouds or aerosol.

The PSC observations are consistent with the retrieved temperature distribution shown in Figure 4.4. Around flight altitude temperatures below the existence temperatures for NAT[2] are found. Towards lower altitudes,

---

[2]T_{NAT} is here calculated using retrieved temperature and HNO_3 from MIPAS-STR and considering H_2O mixing ratios provided by collocated in-situ measurements from the FLASH instrument. The in-situ water vapour measurements from FLASH-A where kindly provided by Sergey Khaykin, Central Aerological Observatory, Dolgoprudny, Moscow region, Russia. Thermodynamic parameters of NAT were taken from Worsnop et al. [1993].

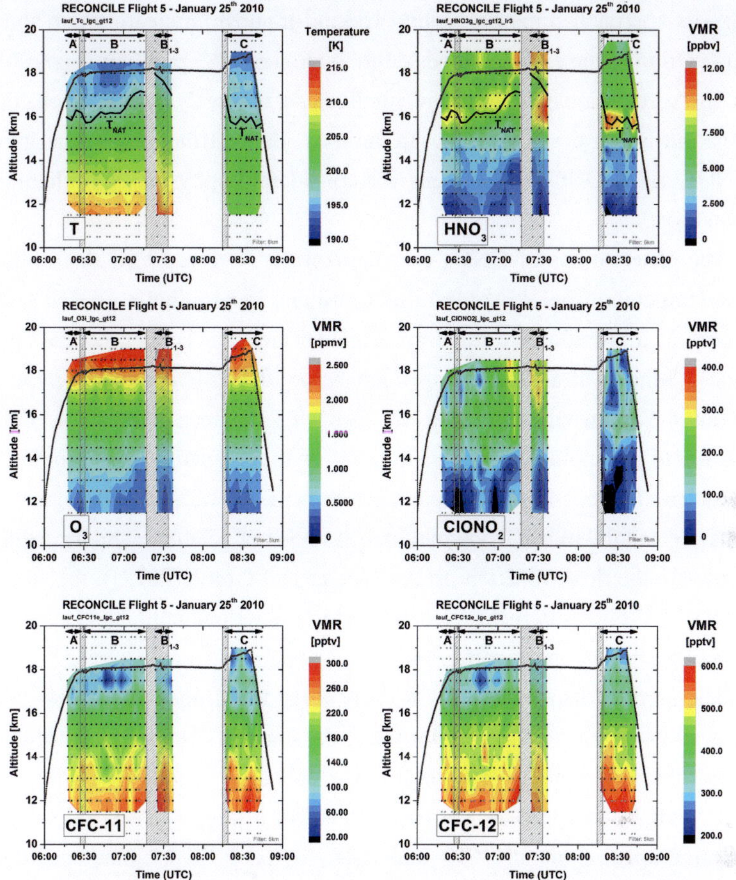

Figure 4.4: Flight altitude and retrieved vertical cross-sections of temperature and HNO$_3$, O$_3$, ClONO$_2$, CFC-11 and CFC-12 mixing ratios for the flight on January 25th 2010. Grey hatched areas marking turns performed by the Geophysica with less reliable pointing information potentially affecting the MIPAS-STR retrievals. Retrieval gridpoints indicated by black/grey dots are filtered for vertical resolutions better than 5 km (6km for temperature) for interpolation (black dots). Typical vertical resolutions are discussed in Section 3.2.5.

T_{NAT} is approached mostly around 16 km (or above) indicating that above these altitudes the existence and sedimentation of NAT particles is possible.

The vertical cross-section of O_3 indicates a relatively homogeneous distribution inside the vortex air. In contrast, the distributions of the gases HNO_3 and $ClONO_2$ show signs for considerable physical and chemical processing.

The observations of HNO_3 complement the observed PSC distribution and temperature field: Around flight altitude, local minima of HNO_3 are found, as a considerable fraction of this species was condensed PSC particles. Mainly at altitudes of $\sim$1 km below the PSC layers indicated in Figure 4.3 and where temperatures approach T_{NAT}, local maxima of HNO_3 are found and point on the evaporation of sedimented HNO_3-containing particles. There is found considerable coincidence between altitudes with $T=T_{NAT}$ and the local HNO_3 maxima. As it is not yet clear which particles dominate the denitrification process (Peter and Grooß [2012]), the shown MIPAS-STR measurements clearly support the sedimentation and evaporation of particles composed of NAT.

The vertical distribution of $ClONO_2$ along flight track shows low concentrations mostly <300 pptv for the entire flight at stratospheric altitudes. This finding is consistent with chlorine activated conditions during this flight: HCl and $ClONO_2$ are transformed to a large fraction into active chlorine species (mainly ClO and ClOOCl) on PSC particles according to the reactions R1–R3. No significant deactivation into $ClONO_2$ is observed. Only around 7:30 UTC slightly increased $ClONO_2$ mixing ratios are found around flight altitude, which is consistent with weaker signatures from PSCs in the MIPAS-STR measurements and warmer temperatures. More gas-phase HNO_3 was available here, probably providing NO_2 via photolysis under the twilight conditions of this flight and therefore allowing early deactivation of ClO into $ClONO_2$ via R8. A study on diurnal variations of reactive chlorine and nitrogen oxides based on measurements of of the ballon-borne instrument MIPAS-B at the day before the discussed flight was performed by Wetzel et al. [2012] and supports the conditions at stratospheric altitudes (i.e. presence of PSCs and chlorine activation) indicated by MIPAS-STR.

While the meteorological analysis suggests that this flight was carried out fully within vortex air, the distributions of CFC-11 and CFC-12 show inhomogeneities around flight altitude, indicating filaments and/or differential descent of air inside the vortex. The relatively long-lived CFCs can be used as tracers and their vertical distributions allow assignment of the corresponding airmasses to vortex-air, extra-vortex air and filaments (compare Section 4.2.4).

In summary, the results from this flight give a consistent picture of the late PSC phase of the Arctic winter 2009/10 and support the presence of PSCs including NAT particles, an ongoing dentrification process and chlorine-activated conditions. The HNO_3 retrieval from the PSC-affected MIPAS-STR measurements (cloud index miniumum values of 1.6 around flight altitude) resolves narrow vertical structures in the order of 1 km, as indicated by the local HNO_3 maxima around 16 km altitude below the PSC clouds.

4.2.2 Flight on January 30th 2010: Developed denitrification in cold vortex after PSC phase

The flight on January 30th was carried out directly after the end of the Arctic winter 2009/10 PSC phase. Practically no more PSCs were observed by CALIPSO at the time of the flight (Pitts et al. [2011]). Although at the beginning of the major stratospheric warming and vortex split, at this stage of the winter the vortex was still compact and cold (Dörnbrack et al. [2012]). Furthermore, the flight was characterised by chlorine activated conditions, as discussed by Sumińska-Ebersoldt et al. [2012].

RECONCILE flight 7 was carried out again from Kiruna (Sweden) in the morning (6:36-10:15 UTC). The flight track and horizontal sampling

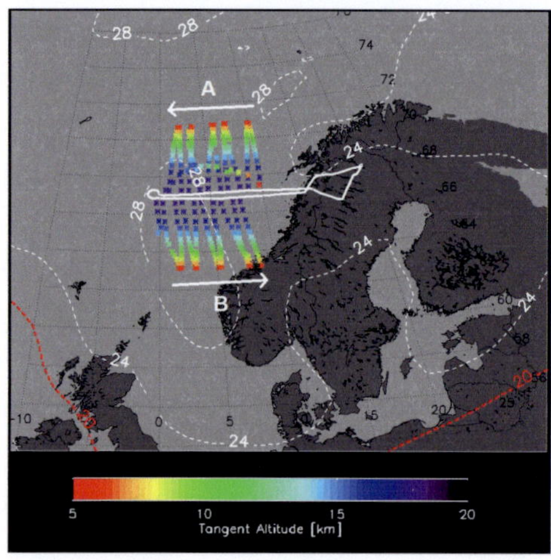

Figure 4.5: Flight track of the Geophysica during RECONCILE flight 7 on January 30th 2010. Tangent points of MIPAS-STR observations. PV-isolines (in PVU) at the potential temperature level of 430 K ($\sim$ 17 km) at 6:00 UTC from ECMWF ERA-Interim reanalysis indicated by dashed white lines. Vortex edge marked by dashed red line according to Nash et al. [1996].

of MIPAS-STR is indicated in Figure 4.5. The PV distribution at 6:00 UTC at about flight altiude indicates that the entire flight was situated well within the polar vortex. Due to electromagnetic compatibility problems only measurements in the time interval between 7:30 to 9:00 UTC were suitable for scientific evaluation.

However, extended areas north of the flight track under night conditions (section A, outbound flight leg from Kiruna) and south of the flight track (section B, inbound flight leg track towards Kiruna) during sunrise were covered by the remaining MIPAS-STR measurements. The flight altitude, vertical distribution of the tangent points and the interpolated distribution of the cloud index are shown in Figure 4.6. The entire flight is characterised by high cloud-index values at stratospheric altitudes indicating the absence of PSCs. Cloud indices around 4 are reached at altitudes $\sim$10 km due to

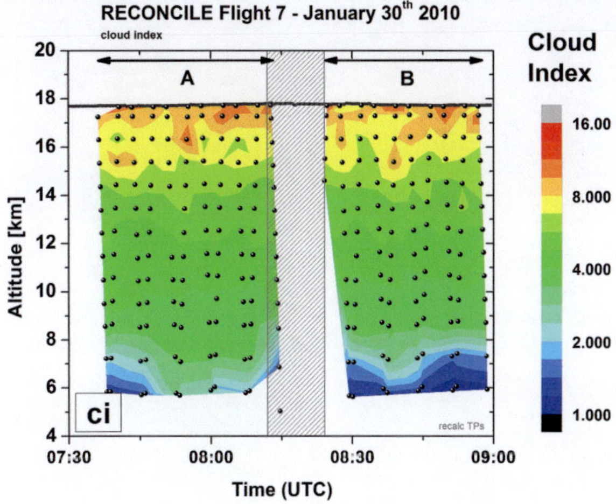

Figure 4.6: Flight altitude, vertical distribution of MIPAS-STR tangent points and cloud index for the flight on January 30th 2010.

increased particle/aerosol loading or optically thin clouds at lower altitudes. Values close to 1 indicate dense tropospheric clouds below 8 km.

The retrieved vertical cross-sections of temperature, O_3, HNO_3, $ClONO_2$ and the CFCs are presented in Figure 4.7. The retrieved temperatures remain above T_{NAT} during the entire flight. The vertical distribution of O_3 is relatively homogeneous along flight track, indicating slightly higher values around flight altitude in the southern area covered by the MIPAS-STR tangent points. In contrast, the vertical distribution of HNO_3 is characterised by considerable HNO_3 redistribution through denitrification: Extended HNO_3 maxima with peak values of 16 ppbv at about 16 km are found for the measurements north (section A) and south (section B) of the flight track.

The Geophysica obviously traversed an extended filament enriched in HNO_3 through renitrification. A dynamic structure can be excluded here,

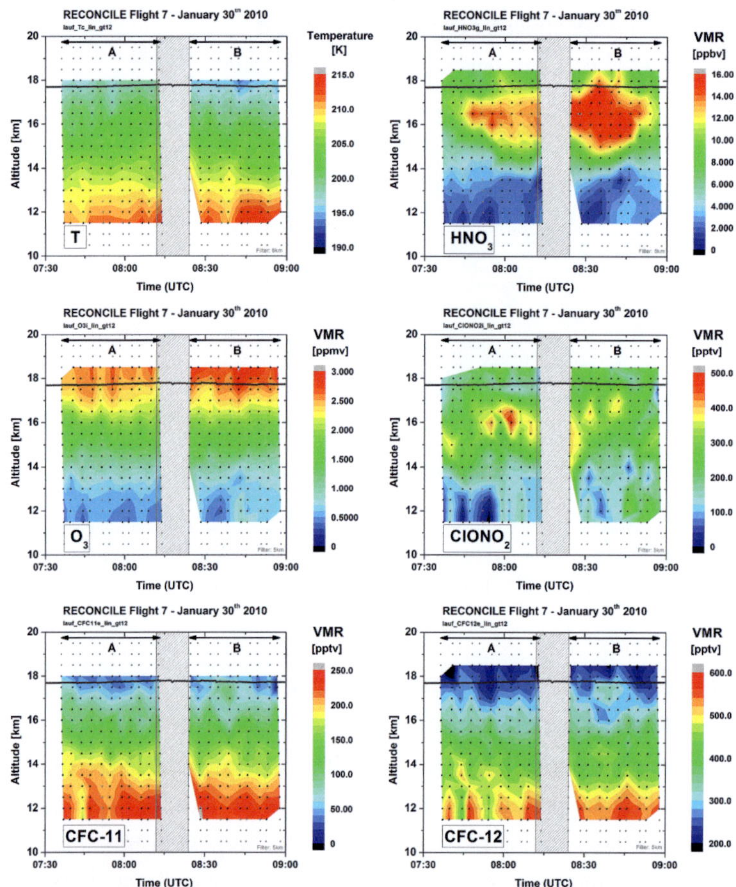

Figure 4.7: Flight altitude and retrieved vertical cross-sections of temperature and HNO$_3$, O$_3$, ClONO$_2$, CFC-11 and CFC-12 mixing ratios for the flight on January 30th 2010. Grey hatched areas marking turns performed by the Geophysica with less reliable pointing information potentially affecting the MIPAS-STR retrievals. Retrieval gridpoints indicated by black/grey dots are filtered for vertical resolutions better than 5 km (6km for temperature) for interpolation (black dots). Typical vertical resolutions are discussed in Section 3.2.5.

as no such structure is identified in the distributions of the tracers CFC-11 and CFC-12 and also not in the O_3 distribution. The vertical cross-section of $ClONO_2$ shows low overall mixing ratios of this species consistent with chlorine-activated conditions. Slightly enhanced $ClONO_2$ mixing ratios are found around 16 km and hint on early chlorine deactivation at these altitudes. This finding is consistent with the excess supply of HNO_3, probably providing limited NO_2 from photolysis and thereby allowing the transformation of ClO into $ClONO_2$ via reaction R8. The distributions of the tracers CFC-11 and CFC-12 show substructures along flight track, hinting on filaments inside the polar vortex.

Accordingly, the MIPAS-STR retrieval results are consistent with conditions free of PSCs. The results support chlorine activation and show considerable patterns of renitrification.

4.2.3 Flight 11 on March 2nd 2010: Aged vortex and filaments

RECONCILE flight 11 was carried out in the late phase of the Arctic winter 2009/10. After the sudden stratospheric warming accompained by a vortex split at the end of January and a continuous warming during February, this flight allowed probing of the late compact vortex remnant at the begin of March (Dörnbrack et al. [2012]). During the flight on March 2nd 2010, the remaining compact vortex lobe was moving from the Canadian sector towards Siberia. The southern part of the vortex remnant passed Spitsbergen in south-west direction and allowed sampling of aged vortex air in the northern part of the flight.

The flight track of the Geophysica is shown together with the MIPAS-STR tangent points in Figure 4.8. As this flight was designated as the second part of a relay flight from Kiruna (Sweden) to Longyearbyen (Norway, Spitsbergen) and back, the Geophysica ascended in the late morning of March 2nd 2010 from Longyearbyen at 9:35. In order to obtain maximum sampling within vortex air, first a north-western flight track was followed.

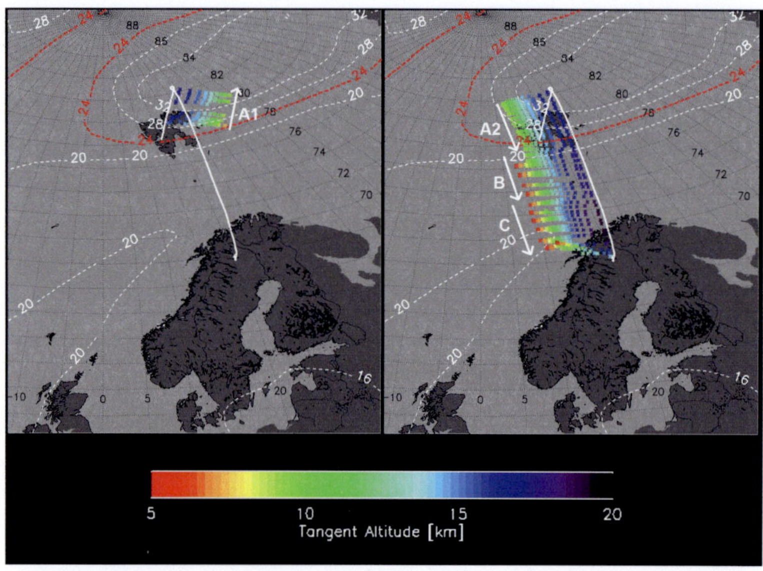

Figure 4.8: Flight track of the Geophysica during RECONCILE flight 11 on March
2nd 2010. Tangent points of MIPAS-STR observations. PV-isolines (in
PVU) at the potential temperature level of 450 K ($\sim$ 17 km) at 12:00
UTC from ECMWF ERA-Interim reanalysis indicated by dashed white
lines. Vortex edge marked by dashed red line according to Nash et al.
[1996].

A turn was performed at $\sim$82°N and then the Geophysica returned towards
Kiruna (landing at 13:35 UTC). The PV-isolines indicate the situation of
the late vortex remnant in the northern part of the flight, with the vortex
edge crossing Spitsbergen at about 78°N.

As discussed in Section 3.2.5, the MIPAS-STR measurements in the
northern part of the flight cover strong PV gradients associated to the vortex
edge, which complicated the discussed in-situ comparisons. In the southern part of the flight the 20 PVU-line north-west of Scandinavia hints on the
presence of a filament in south-west to north-east direction at $\sim$17 km altitude. This structure probably corresponds with the systematics found in the

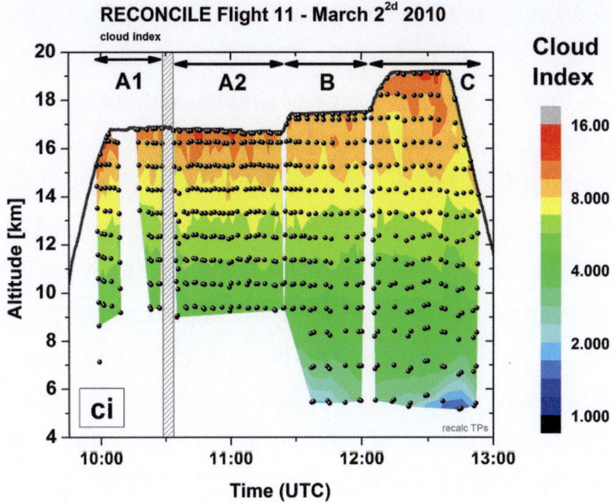

Figure 4.9: Flight altitude, vertical distribution of MIPAS-STR tangent points and cloud index for the flight on March 2nd 2010.

in-situ comparisons for the CFCs in Section 3.2.5[3]. The vertical distribution of the MIPAS-STR tangent points is shown in Figure 4.9 together with the interpolated cloud index. At stratospheric altitudes high cloud indices are found for the entire flight and indicate cloud-free conditions. As for the flight on January 30th, the cloud index approaches 4 at altitudes below 10 km as a consequence of increased particle/aerosol loading or thin clouds. Lowest values close to 1 indicate dense clouds only at the very end of the flight around 12:45 UTC (section C) at altitudes lower than 6 km. In order to enhance the horizontal sampling in the vortex air, in the first part of the flight the modified measurement scenario omitting tangent altitudes below 9 km and with reduced upward viewing measurements was performed, as reflected in the dense horizontal distribution of the tangent points.

---

[3]I.e. MIPAS-STR measurements indicating lower mixing ratios of the CFCs compared to the in-situ measurements due to integrating radiation contributions from this filament.

In Figure 4.10 the vertical cross-sections of temperature, O_3, HNO_3, $ClONO_2$ and the PSCs are presented. Colder temperatures are found for the first part of the flight for the highest latitudes and within the late vortex (prior to 11:00 UTC in sections A1 and A2), however well above the existence temperatures of PSCs. The distribution of O_3 shows low mixing ratios inside the late compact vortex remnant (vortex edge passed at $\sim$11:15 UTC) characteristic for vortex air. In the latter flight part, higher mixing ratios of O_3 are found, as extra-vortex air was sampled, while the observed inhomogeneities in the vertical O_3 distribution along flight track indicate airmasses and filaments of different origin.

Differences between vortex and extra-vortex air are identified in the vertical cross-section of HNO_3: Between 10:00 and 11:15 UTC the late vortex remnant above and north of Spitsbergen becomes clearly visible above 14.5 km. Around flight altitude low HNO_3 mixing ratios of about 5 to 6 ppbv hint on still denitrified air. An underlying sharp potential renitrification layer centered at 15.5 km is identified. Peak mixing ratios of 8 ppbv are observed at a resolved vertical width of about 1 km. A structure of extra-vortex air is found between 11:35 and 12:00 UTC (section B), as indicated by low HNO_3 mixing ratios typical for air from mid- and low latitudes.

The structure is diagonally linked to further areas of extra-vortex air in the last part of the flight. Two overlaying filaments characterised by enhanced HNO_3 and separated by air with low HNO_3 mixing ratios are found in the last flight section C between 12:05 to 12:45. The filament centered at 14.5 km with a vertical extension $\sim$3 km shows strongly enhanced HNO_3 mixing ratios of up to 10 ppbv and is probably a renitrification structure originating from the polar vortex. The more narrow maximum above with a vertical extension of $\sim$1.5 km also shows enhanced HNO_3 mixing ratios up of to 9 ppbv and also hints on vortex air. However, tracer analyses and a model study by Kalicinsky [2013] showed that this filament has low vortex character. The structures observed by MIPAS-STR during this flight are confirmed comparisons with the infrared limb-sounder CRISTA-NF for the

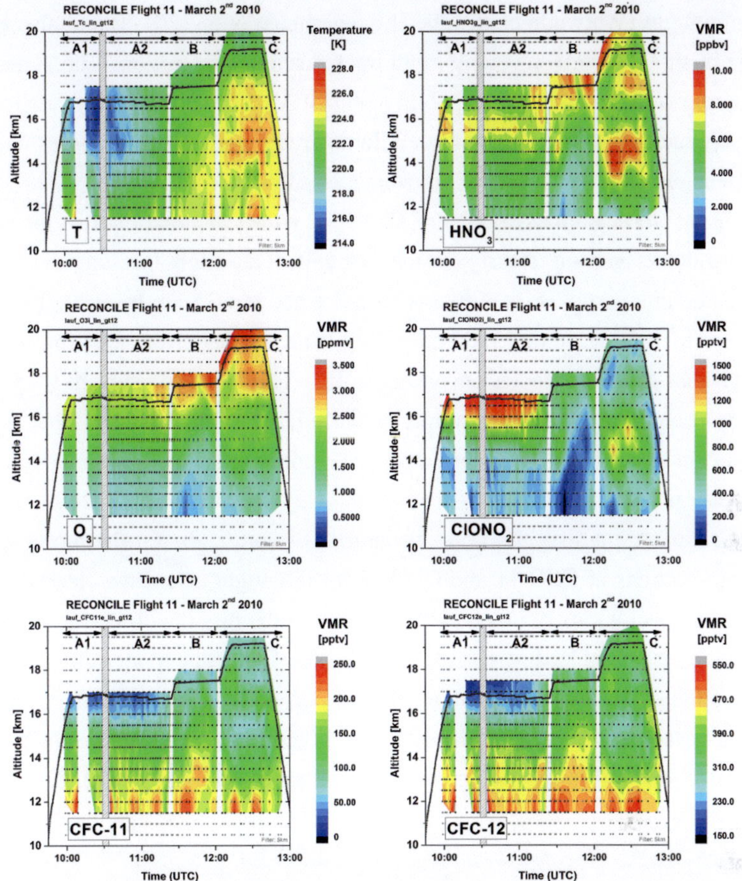

Figure 4.10: Flight altitude and retrieved vertical cross-sections of temperature and
HNO$_3$, O$_3$, ClONO$_2$, CFC-11 and CFC-12 mixing ratios for the flight
on March 2nd 2010. Grey hatched areas marking turns performed by
the Geophysica with less reliable pointing information potentially affecting the MIPAS-STR retrievals. Retrieval gridpoints indicated by
black/grey dots are filtered for vertical resolutions better than 5 km for
interpolation (black dots). Typical vertical resolutions are discussed in
Section 3.2.5.

same flight (Woiwode et al. [2012]; Ungermann et al. [2012]) and also by the in-situ comparisons along flight track discussed in section 3.2.5. These comparisons confirm the capability of MIPAS-STR of resolving vertical fine-structures in the order of one kilometer and horizontal structures with extensions of a few tens of kilometers along flight track.

In the vertical cross-section of $ClONO_2$, vortex and extra vortex air are separated even more clearly: In the first half of the flight (sections A1 and A2) the late vortex is identified at altitudes above $\sim$15 km by strongly enhanced $ClONO_2$ mixing ratios resulting from the deactivation of formerly activated chlorine species according to R8. The lower filament in section C is also characterised by strongly enhanced $ClONO_2$, supporting that this filament stems from the polar vortex. The upper structure is separated less clearly in the $ClONO_2$ cross-section, but slightly enhanced $ClONO_2$ mixing ratios might also hint on contributions of vortex air. Furthermore, the cross-sections of CFC-11 and CFC-12 for this flight also show clearly the late compact vortex remnant in the first half of the flight, with vortex air being characterised by low mixing ratios of the CFCs due to air subsidence. Also the overlaying filaments in the last part of the flight (section C) appear faintly in the cross-sections of the CFCs, but show much weaker contrasts compared to the background mixing ratios compared to HNO_3 and $ClONO_2$.

The MIPAS-STR retrieval results from the flight on March 2nd 2010 provide a consistent picture of the remaining aged vortex lobe after the sudden stratospheric warming and vortex split as well as of the overall dynamical situation: Remnants of de- and renitrification layers are observed inside and below the vortex remnant and in filaments. Strongly enhanced $ClONO_2$ in vortex air indicates a high degree of chlorine deactivation. The retrieved vertical cross-sections demonstrate that narrow vertical and horizontal structures can be well resolved by this measurement technique.

4.2.4 Discussion: Denitrification and chlorine deactivation in the Arctic winter 2009/10

In this section the MIPAS-STR retrieval results of HNO_3 and $ClONO_2$ of the three discussed flights during the RECONCILE period are brought into a quantitative context to study denitrification and chlorine deactivation. The flights cover the end of the cold phase of the Arctic winter 2009/10 at the end of January 2010 and the late phase of the polar vortex about one month later after a sudden stratospheric warming that resulted in a vortex split. The MIPAS-STR results from the individual flights show the following atmospheric conditions:

- **January 25th 2010**: Wide-spread PSCs, temperatures below T_{NAT}, probably ongoing denitrification and chlorine activation are observed. Local HNO_3 maxima at altitudes with $T \approx T_{NAT}$ below PSC clouds point on sedimentation and evaporation of HNO_3-containing particles consisting of NAT.

- **January 30th 2010**: Conditions free of PSCs and a wide-spread developed renitrification layer indicated by HNO_3 maxima around 16 km altitude are found. Locally enhanced $ClONO_2$ mixing ratios inside the HNO_3 maxima indicate early chlorine deactivation according to R8, probably involving NO_2 resulting from photolysis of excess-HNO_3.

- **March 2nd 2010**: The observations show the late vortex, extra-vortex air and filaments. Denitrified air is still present in the vortex air and HNO_3 maxima due to renitrification are found below the vortex remnant and in filaments. Strongly enhanced $ClONO_2$ mixing ratios inside the vortex and the large lower filament in the end of the flight indicate chlorine-deactivated air.

The retrieved mixing ratios from the different flights are compared via their correlations with the long-lived tracer CFC-12. Greenblatt et al.

[2002] showed that vortex air can be accurately identified by in-situ measurements of a chemical inert tracer such as N_2O versus potential temperature. Vortex air is characterised by strong downwelling through diabatic cooling within the polar vortex, and the grade of downwelling is reflected by the vertical displacement of levels with certain tracer mixing ratios. Important preconditions are that the tracer shows a sufficiently strong gradient versus altitude and that the mixing ratios do not drop to zero in the considered vertical range.

From the trace gases discussed in this work, CFC-12 with an atmospheric lifetime of $\sim$100 years at tropospheric altitudes (Montzka et al. [2011]) is suited best for this approach, showing a sufficiently strong gradient and sufficiently high mixing ratios. Furthermore, the retrieved mixing ratios of CFC-12 are less affected by horizontal gradients as in the case of CFC-11 (i.e. at the vortex edge or in the presence of filaments along the viewing directions, compare Section 3.2.5). Figure 4.11 shows retrieved CFC-12 mixing ratios plotted versus potential temperature (derived from retrieved temperature and pressure profiles from the ECMWF analysis used for the retrieval) for the discussed flights. Data points labeled with 'vortex' correspond to measurements inside the polar vortex, as seen in the PV distributions for the individual flights (previous sections) and indicated by the vertical cross-sections of the tracers CFC-11 and CFC-12.

The flights on January 25th and 30th were both carried out fully inside the polar vortex (compare Sections 4.2.1 to 4.2.3, PV distributions and tracer fields). However, the scatter of the MIPAS-STR data points and the structures seen in the vertical cross-sections of CFC-11 and CFC-12 in Figures 4.4 and 4.7 indicate influence from air with less vortex-character on the MIPAS-STR measurements.

The flight on March 2nd was carried out in in the late vortex remnant and also extra-vortex air from different origin. From this flight only data points between begin of the flight (first MIPAS-STR scan at 9:58) UTC and 11:00 UTC are attributed to 'vortex'. For this part of the data the vortex

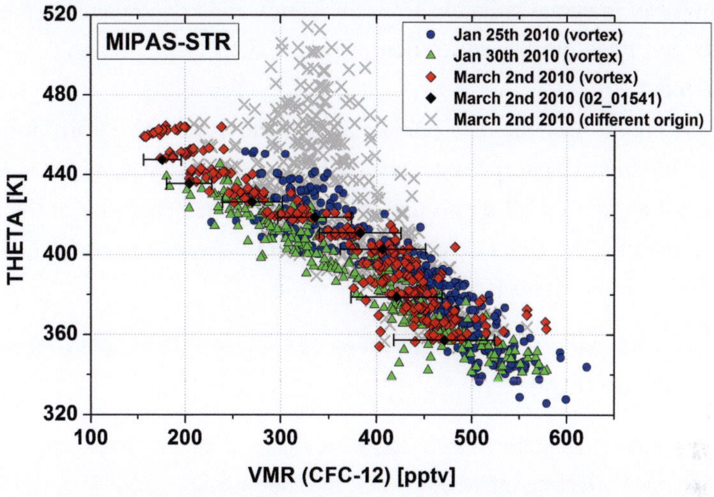

Figure 4.11: Retrieved mixing ratios of CFC-12 versus potential temperature for the indicated flights. Flights on January 25th and 30th: All measurements attributed to 'vortex', compare PV distributions Figures 4.1 and 4.5. Flight on March 2nd: Measurements prior to 11:00 UTC attributed to 'vortex', compare Figures 4.8 (PV distribution) and 3.18 (tracers CFC-11 and CFC-12 along flight track). Measurements after 11:00 UTC attributed to 'different origin'. Selected data points with error bars indicated for scan 02_01541 on March 2nd with deepest penetration of the vortex (compare Figure 4.8, northernmost scan in section A1). Vertical error bars for potential temperature within sizes of data points.

character of the associated airmasses is indicated by the PV field derived from ECMWF (Figure 4.8) and the retrieved distributions of the tracers CFC-11 and CFC-12 (Figure 4.10). Here, $ClONO_2$ also is a tracer for vortex air, since strongly enhanced mixing ratios of this gas indicate chlorine-deactivated air. Measurements after 11:00 UTC however situated within the vortex according to the PV distribution (compare Figure 4.8) are conservatively attributed to 'different origin', as the in-situ comparisons (compare

Figure 3.18) indicate that the corresponding MIPAS-STR measurements probably integrate radiation from the vortex edge region/outside the vortex and therefore the correlation of potential temperature and CFC-12 is affected.

In Figure 4.11 the data points attributed to vortex air clearly follow a similar approximately linear relationship for CFC-12 mixing ratios lower than 400 pptv for all flights and confirm the vortex character of the measurements. The slight vertical displacement between the correlations for the individual flights is attributed to:

- different grades of downwelling of the vortex air sampled during different flights

- inhomogeneities inside the vortex (i.e. filaments from extra-vortex air) affecting the MIPAS-STR measurements

- contributions from radiation from the vortex edge (i.e. influence of horizontal tracer contrasts and gradients along the viewing direction, especially for the flight on March 2nd)

- potential mixing of vortex air with extra-vortex air during the polar winter and transport through the vortex edge (compare Grooß et al. [2008]).

Therefore, it is emphasised that the measurements at the end of January and begin of March show different phases of the polar vortex 2009/2010, which are characterised by different grades of downwelling of the airmasses, dynamic disturbances, and mixing with extra-vortex air, which has to be reminded when interpreting the following comparisons.

The data points corresponding to 'different origin' air and filaments show a significant portion with higher mixing ratios, leaving the characteristic relationship at theta levels higher than 400 K (corresponding to $\sim$14 km altitude on March 2nd), resulting in the typical 'Y-shape' of this plot. This fraction of data points indicates extra-vortex air characterised by higher

CFC-12 mixing ratios compared to the lower CFC-12 mixing ratios inside the vortex as a consequence of air subsidence.

The scatter in the MIPAS-STR data points is much higher than in the results from more precise in-situ measurements discussed by Greenblatt et al. [2002] due to higher uncertainties in the MIPAS-STR measurements. Furthermore, effects from horizontal gradients and contrasts in the CFC-12 mixing ratios and vertical smoothing by the retrieval 'washes out' a sharper contrast here. However, the shown result confirms the assignment of 'vortex' air performed from the PV distributions and the retrieved tracer distributions of CFC-11 and CFC-12 and demonstrate that distinguishing between vortex and extra-vortex air is possible in the MIPAS-STR measurements.

Based on the assignment of 'vortex', the correlations of HNO_3 and $ClONO_2$ versus CFC-12 derived from the MIPAS-STR measurements are shown in Figures 4.12 and 4.13. For comparison with the MIPAS-STR results for $ClONO_2$ also the correlations of ClO under sunlight conditions (solar zenith angles $<90°$) and CFC-12 derived from the in-situ instruments HALOX (HALogen OXide monitor, Sumińska-Ebersoldt et al. [2012]) and HAGAR (see section 3.2.5) aboard the Geophyisca for the flights on January 30th and March 2nd are shown[4]. From the correlations versus CFC-12 the changes in the specific trace gas mixing ratios due to chemical and physical processing are become visible:

Denitrification

Significant HNO_3 redistribution is observed for the flights in January, as seen in the very different correlations of HNO_3 and CFC-12 for the individual flights. For the flight on January 25th, HNO_3 mixing ratios scattering

---

[4]In-situ measurements of daytime ClO and CFC-12 profiles inside vortex air are considered for January 30th in the time interval 9:30 to 9:50 UTC and for March 2nd in the time interval 9:45 to 11:00. HALOX ClO data: Courtesy Fred Stroh, Research Centre Jülich GmbH, Germany; HAGAR CFC-12 data: Courtesy Michael Volk, University of Wuppertal, Germany.

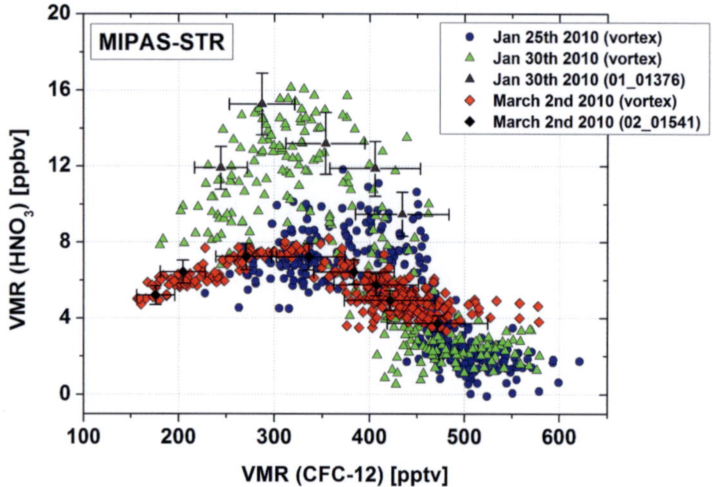

Figure 4.12: MIPAS-STR correlations of HNO$_3$ versus CFC-12 for data points attributed to 'vortex' showing vertical HNO$_3$ redistribution through de-/renitrification. Selected data points with error bars indicated for scan 02_01376 on January 30th (compare Figure 4.5, third scan in section B) and for scan 02_01541 on March 2nd with deepest penetration of the vortex (compare Figure 4.8, northernmost scan in section A1).

to lowest values of about 4 ppbv at CFC-12 mixing ratios around 300 pptv (highest altitudes around 18 to 19 km probed during this flight) point on HNO$_3$ being condensed in PSC particles. This is consistent with temperatures below T$_{NAT}$ around flight altitude during this flight (see section 4.2.1). A significant portion of data points scattering towards higher HNO$_3$ mixing ratios for CFC-12 mixing ratios between 350 and 450 pptv compared to the 'underlying' correlation indicated by the measurements from all flights indicates the early formation of renitrification layers blow the PSCs at this stage of the winter.

During the flight on January 30th, five days later and characterised by no more PSC occurrence, the entire HNO_3 previously condensed in PSC particles was evaporated. An extended renitrification layer is indicated by the strongly enhanced maximum mixing ratios of HNO_3 corresponding with CFC-12 mixing ratios between 300 to 350 pptv. Maximum HNO_3 mixing ratios increased by up to 9 pptv are found compared to the correlations for the other flights.

For the flight on March 2nd only weak signs for renitrification remnants are indicated by the correlation, with only very few data points scattering to higher HNO_3 mixing ratios compared to the underlying correlation indicated by the measurements from all shown flights. The characteristic decrease in the HNO_3 mixing ratios especially for this flight for lowest CFC-12 mixing ratios around 200 pptv (i.e. altitudes around 17 km) hints on still denitrified air in the late vortex.

The measured vertical distribution of HNO_3 is clearly affected by HNO_3 redistribution through denitrification in the Arctic winter 2009/10. The observed patterns during the individual flights (see Sections 4.2.1-4.2.3) are locally variable and can be seen as 'fingerprints' from the denitrification process. In Section 4.3 the characteristic vertical HNO_3 distributions derived for the January flights are used for a detailed study with the CLaMS model considering denitrification by NAT particles with reduced settling velocities. The study aims on explaining the size distributions of large potentially HNO_3-containing particles indicated by in-situ measurements during RECONCILE (von Hobe et al. [2012]).

Chlorine Deactivation

The correlations of $ClONO_2$ and ClO versus CFC-12 show a complementary picture for CFC-12 mixing ratios lower than 400 pptv (i.e. above $\sim$14 km at the flight on March 2nd): In January, strongly enhanced mixing ratios of ClO and lowest $ClONO_2$ mixing ratios point on chlorine activation (see also Sumińska-Ebersoldt et al. [2012]). Under daytime conditions the equilibrium between ClOOCl and ClO (R4 to R7) is practically completely shifted to ClO. Therefore the HALOX ClO measurements can be seen as an approximate measure of entire active chlorine. In March the situation is vice versa: Lowest ClO mixing ratios are observed in the late vortex air, while $ClONO_2$ is strongly enhanced as the consequence of deactivation of active chlorine into this species via R8.

From the MIPAS-STR measurements in the mixing ratio bin between 240 and 260 pptv CFC-12 it is estimated that the average mixing ratio of $ClONO_2$ in this interval on March 2nd is enhanced by 1133 ($\pm$295) pptv compared to January 30th. The indicated maximum error for MIPAS-STR considers the combined 1σ-error (Section 3.2.4) errors of the individual $ClONO_2$ retrieval results as an estimate for systematic errors. This is motivated by the fact that the in-situ comparisons (Section 3.2.5) show that the MIPAS-STR measurements mostly agree with the collocated in-situ measurements within the combined 1σ-error. Therefore, taking into account the characteristic combined 1σ-error of the single $ClONO_2$ data points as an conservative estimate for systematic errors in the averaged measurements seems appropriate.

Furthermore it was estimated, how systematic errors in the retrieved CFC-12 mixing ratios affect the comparison. Offsets in the retrieved CFC-12 mixing ratios corresponding to the combined 1σ-error for CFC12 were assumed and the resulting differences in the corresponding $ClONO_2$ mixing ratios were estimated.

Finally, the discussed combined 1σ-errors for the $ClONO_2$ mixing ratios and the maximum offsets in $ClONO_2$ resulting from the potential systematic errors in the CFC-12 mixing ratios for each correlation were added to yield the indicated maximum error for the difference in $ClONO_2$. Thereby, the MIPAS-STR errors are considered to have predominantly systematic character, resulting in an conservative error estimation. However, this error estimation for the difference in $ClONO_2$ between the two flights allows for a direct and meaningful comparison with the in-situ data from HALOX also

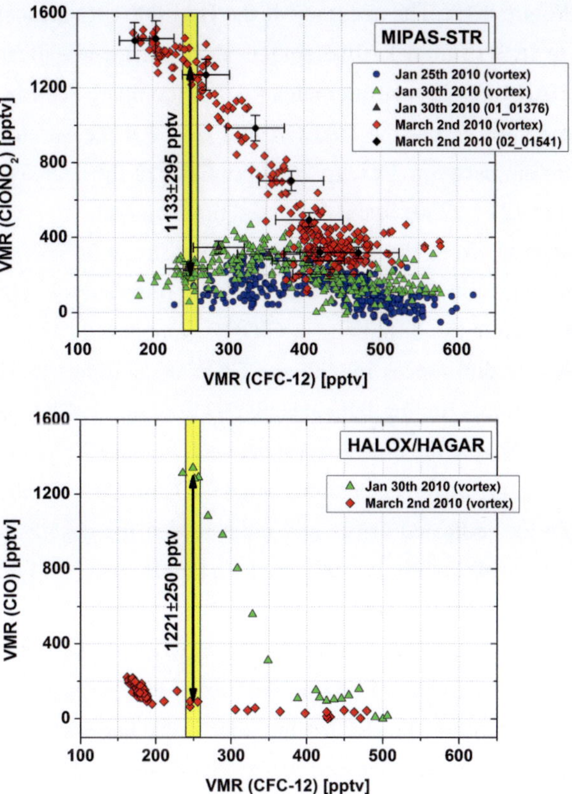

Figure 4.13: Upper panel: MIPAS-STR correlations of ClONO$_2$ versus CFC-12 for data points attributed to 'vortex', showing deactivation of active chlorine into ClONO$_2$ between end of January and begin of March 2010. Selected data points with error bars indicated for scan 02_01376 on January 30th (compare Figure 4.5, third scan in section B) and for scan 02_01541 on March 2nd with deepest penetration of the vortex (compare Figure 4.8, northernmost scan in section A1). Lower panel: Correlations of ClO versus CFC-12 from in-situ measurements aboard the Geophysica during the flights on January 30th and March 2nd 2010 under sunlight conditions. Vortex air here identified from PV distributions in Figures 4.5 and 4.8 and HAGAR tracer measurements.

shown in Figure 4.13. The accuracy of the HALOX ClO-measurements is estimated to 18 % (Sumińska-Ebersoldt et al. [2012]), while the uncertainty of the HAGAR CFC-12 measurements ≤ 1.7 % (compare Section 3.5.2) can be neglected here. From the HALOX and HAGAR measurements in the mixing ratio bin between 240 and 260 pptv CFC-12 the decrease in ClO is estimated to 1221 ($\pm$250) in the discussed time interval.

To summarise, from the shown correlations it can be seen that in the CFC-12 mixing ratio bin 240 to 260 pptv the difference in ClO is 1221 ($\pm$250) pptv, while the difference in $ClONO_2$ amounts 1133 ($\pm$295) pptv. Considering the differences in ClO and $ClONO_2$ as representative for the polar vortex air, this finding indicates that 93 % of active ClO were deactivated into $ClONO_2$ in the time interval between January 30th and March 2nd 2010. Thereby, the differences in measured ClO and $ClONO_2$ are in agreement within the indicated errors and indicate that the deactivation of active chlorine almost completely occurred via the fast reaction into $ClONO_2$ (R8).

It is pointed out that here different phases of the polar vortex characterised by different extent of mixing with extra-vortex air are considered, which might affect the CFC-12 mixing ratios in the late vortex and the vortex boundary region sampled during the flight on March 2nd. However, taking into account potential effects from in-mixing of extra-vortex air, the shown results clearly indicate that the chlorine deactivation predominantly occurred into $ClONO_2$.

The shown results indicate that despite the unusually cold phase of this Arctic polar winter accompanied by significant denitrifcation still enough NO_2 was available from HNO_3-photolysis at the corresponding altitudes, allowing an effective deactivation of active chlorine into $ClONO_2$.

Further studies on deactivation of active chlorine into $ClONO_2$ based on balloon-borne and satellite-borne limb sounding measurements are discussed by von Clarmann et al. [1997], Dufour et al. [2006], Hayashida et al. [2007] and Wetzel et al. [2010] and largely confirm the efficiency of this important pathway of chlorine deactivation in Arctic winters. Furthermore, Konopka et al. [2003] report a combined study with the CLaMS model and in-situ measurements on chlorine deactivation in the Arctic winter 2000 characterised by low effects from mixing with extra-vortex air. These authors also found comprehensive deactivation of active chlorine into $ClONO_2$ at the end of the winter.

The measurements from for the Arctic winter 2009/10 presented here however correspond to a vortex which split up two times and was mixed with extra-vortex air (Dörnbrack et al. [2012]). Therefore, the shown results might be used for further studies on the impact of mixing with extra-vortex air on chlorine deactivation. Thereby, the high accuracy of the presented MIPAS-STR $ClONO_2$ measurements and the high vertical as well as horizontal resolution along flight track allow accurate comparisons with high-resolution model simulations. Furthermore, the reliable identification of vortex air with MIPAS-STR using the tracer CFC-12 as well as the combination with the collocated HALOX ClO and HAGAR CFC-12 measurements provide a reliable basis for potential further studies with chemistry models.

4.3 Combined study on the denitrification by potential large-dimension NAT particles based on CLaMS simulations and MIPAS-STR measurements

4.3.1 The process of denitrification, open issues and approach for the study

As introduced at the begin of this work, the process of denitrification has important influence on Arctic ozone depletion chemistry (Solomon [1999]). The freezing out of HNO_3-containing hydrates at stratospheric levels above $\sim$18 km leads to the growth of micron sized particles. Most likely these particles are crystalline and composed of NAT, whereas also further metastable HNO_3-containing hydrates are discussed (Peter and Grooß [2012] and references therein). The sedimentation of these particles and evaporation at lower stratospheric altitudes results in an irreversible vertical redistribution of HNO_3. The resulting HNO_3 deficit at higher stratospheric altitudes limits the fast deactivation of ozone-destroying catalytic substances such as ClO by limiting the availability of NO_2 produced from HNO_3 photolysis. Simultaneuosly, these particles sedimented to lower stratospheric altitudes locally enhance the reactive surface capable of chlorine activation. The fact that these particles can exist at warmer temperatures than ice particles and STS droplets further enhances their importance in Arctic stratospheric chlorine chemistry.

Field measurements and laboratory studies have shown that the particles involved in denitrification are likely composed of NAT (Hanson and Mauersberger [1988];Voigt et al. [2000]; Peter and Grooß [2012] and references therein). Further possible candidates are the metastable nitric acid dihydrate (NAD) (i.e. Worsnop et al. [1993]) and potential higher metastable hydrates of HNO_3 (i.e. Marti and Mauersberger [1994] and Tabazadeh and Toon [1996]), though experimental evidence for such particles in the stratosphere is sparse or still missing, respectively. Large HNO_3-containing particles with significant potential for denitrification were first reported by

Fahey et al. [2001]. While the denitrification process is understood qualitatively, a detailed understanding is lacking for several aspects of this important process, complicating quantitative simulations of denitrification (compare Peter and Grooß [2012]):

- The nucleation process of PSC particles. In particular also the nucleation mechanism preferring growth of low number density large HNO_3-containing particles (i.e. NAT) rather than smaller particles with higher number density.

- The nucleation rate of potential NAT particles involved in denitrification.

- The potential role of HNO_3-containing hydrates other than NAT in dentrification.

- The shape and morphology of large HNO_3-containing particles involved in dentrification.

During RECONCILE, rather large potential HNO_3-containing particles were observed by optical in-situ instruments (von Hobe et al. [2012]) with maximum sizes (in diameter) considerably larger than the particles reported by Fahey et al. [2001]. These particles can hardly be reconciled with current theory on growth and nucleation of NAT particles when assuming compact spherical particles and the mass density given by Hanson and Mauersberger [1988].

This study adresses the impact of reduced settling velocities of potential NAT particles on denitrification. The study is motivated by the hypothesis that the large particles detected during RECONCILE (i) have significantly aspheric shape or (ii) are consisting of loosely packed aggregates ('flakes'). Such particle a shape or morphology would result in reduced particle settling velocities due to larger surface areas and might furthermore explain the large particle sizes derived from the in-situ measurements during RECONCILE.

The assumption of significantly aspheric particles is supported by the study of Grothe et al. [2006], who analysed the growth and shape of micron sized NAT particles in the presence and absence of ice domains under laboratory conditions. The authors obtained platelet-shaped particles with diameters in the order of microns in the absence of ice, whereas needles were found in the the presence of larger ice domains. Furthermore, Wagner et al. [2005] found micron sized NAD particles under simulated stratospheric conditions in the cloud chamber AIDA (Aerosol Interactions and Dynamics in the Atmosphere), and their measurements are best explained by assuming significantly oblate particles. Thus, particles involved in denitrification might be (partially) composed of such metastable NAD particles. Furthermore, such particles might be converted into NAT subsequently, with oblate NAD particles acting as template for oblate/platelet-shaped NAT particles.

The assumption of 'flake-like' particles with reduced particle mass density is motivated by the results from Keyser and Leu [1993], who characterised films of HNO_3-containing particles near the composition of NAT under laboratory conditions and reported granular particles forming micron-sized aggregates. Accordingly, NAT particles in the stratosphere might consist of loosely packed aggregates of smaller subunits and might have an approximately spherical net shape.

In the following sections, different scenarios from the CLaMS model (Grooß et al. [2005] and references therein) considering denitrification by NAT particles with reduced settling velocities are analysed[5]. Thereby, sedimentation properties of micron sized ice particles discussed by Westbrook [2008] (references therein) were considered for modelling the sedimentation of potential NAT particles in the Arctic stratosphere. The modelled vertical redistribution of HNO_3 is compared with MIPAS-STR measurements of this species during the RECONCILE flights on January 25th and

---

[5]The specific CLaMS simulations and the particle backward trajectories discussed in the following were kindly provided by Jens-Uwe Grooß, Research Centre Jülich GmbH, Germany.

January 30th 2010 under the conditions of synoptic scale PSCs and under cloud-free conditions, respectively. Both flights were characterised by considerable HNO_3 redistribution. Furthermore, during the flight on January 25th, significant amounts of HNO_3 were condensed in PSC particles, allowing sensitive comparisons of the MIPAS-STR measurements and the CLaMS simulations. The comparison of modelled and measured vertical HNO_3 redistribution provides information on possible properties of the involved particles (i.e. potential shape and morphology) in the context of current theory of dentrification.

The CLaMS model and the denitrification scheme applied in CLaMS are introduced in the following subsection. Then, optical particle in-situ measurements of large-dimension potentially HNO_3-containing particles during the RECONCILE flight on January 25th are discussed. For the positions and times of the in-situ measurements particle backward trajectories for NAT particles modelled with CLaMS were extracted. The trajectories show that the particles measured in-situ cannot be explained by compact spherical NAT particles due to insufficient growing time with $T < T_{NAT}$. Then, modelled and measured HNO_3 redistribition are compared considering CLaMS scenarios with different particle settling velocities. Finally, by comparing measured and simulated denitrification, the consequences for the properties of potential NAT particles measured in-situ during RECONCILE are discussed.

4.3.2 Introduction of the CLaMS model

This subsection briefly introduces the chemical transport model CLaMS (Chemical Lagrangian Model of the Stratosphere). The model is described in more detail by Grooß et al. [2005] (and references therein). CLaMS provides full 3-dimensional modelling of stratospheric chemistry and particle sedimentation. The CLaMS model is driven by background temperature and wind fields. For the simulations in this context these fields were taken from the ECMWF ERA-Interim Reanalysis (see http://www.ecmwf.int/products/data/archive/descriptions/ei/index.html). CLaMS simulates the growth and sedimentation of large NAT particles according to the the scheme of Carslaw et al. [2002]. Horizontal advection of the modelled particles is simulated considering the background wind fields. The sedimentation of the particles is simulated considering gravitational settling and the viscosity of stratospheric air. Thereby, changes in the settling velocities depending on particle growth are taken into account. Another mode of small NAT particles with high number density reported by Grooß et al. [2005] is not considered here.

The CLaMS simulations discussed in the following have a vertical resolution of 0.5 to 0.6 km in the considered altitude range between 20 to 15 km and a horizontal resolution of $\sim$70 km. Furthermore, a constant volume averaged and temperature independent nucleation rate for NAT of $8.0 \cdot 10^{-5}$ cm^{-3} h^{-1} was applied.

In the following subsections, the reference scenario considers denitrification by compact spherical particles with the particle mass density of 1.6 g cm^{-3} described by Hanson and Mauersberger [1988]. Sensitivity runs were carried out considering reduced particle settling speeds to simulate sedimentation of aspheric particles or particles with reduced mass densities and are discussed in more detail in the corresponding subsections.

4.3.3 In-situ measurements of large-dimension potentially HNO_3-containing particles on January 25th 2010 and comparison with CLaMS particle backward trajectories

During RECONCILE dedicated in-situ instruments were deployed aboard the Geophyisca and provided information on stratospheric cloud particles and aerosol. The FSSP-100 instrument (de Reus et al. [2009]) performed measurements of particles with sizes between 1 to 40 μm based on the detection of single scattered laser light. Usually the FSSP-100 measure-

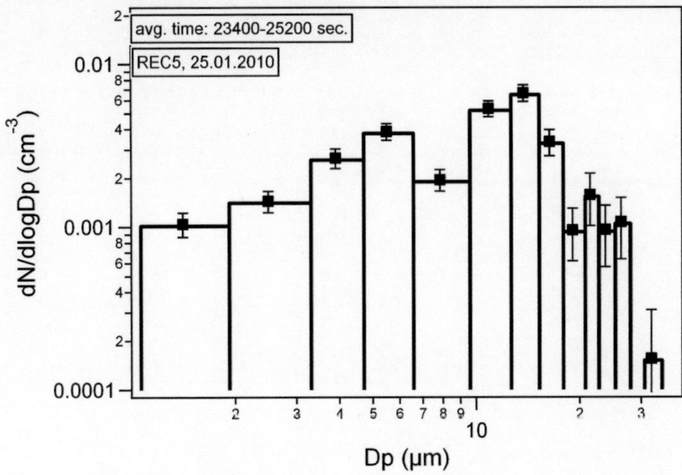

Figure 4.14: Size distribution (in diameter) for particles measured by the FSSP-100 during the RECONCILE flight on January 25th 2010 (Courtesy Sergey Molleker, MPI Mainz, Germany). Measurements correspond to the time interval 6:30 to 7:00 UTC (flight altitude 18 km). N = particle number denstity and Dp = particle diameter.

ments are evaluated assuming spherical particles considering the Mie theory (Mie [1908]) for deriving particle size distributions. For randomly oriented slightly aspheric particles this approach allows deriving meaningful size distributions. Advanced T-matrix calculations allow also deriving sizes for aspheric particles from FSSP measurements, but a priori knowlege on particle shape is needed (Borrmann et al. [2000]).

In Figure 4.14 a particle size distribution derived from the FSSP-100 measurements during the flight on January 25th 2010 in the 6:30 to 7:00 UTC interval is shown. For this analysis spherical particles were assumed. The size distribution shows two main modes of particles with sizes in diameters around 5.5 and 14 μm with considerably high number densities of $\sim$0.004 and 0.007$\sim$ cm^{-3} and further weaker modes. Furthermore, extremely large particles with diameters of $>$30 μm are indicated, exceeding

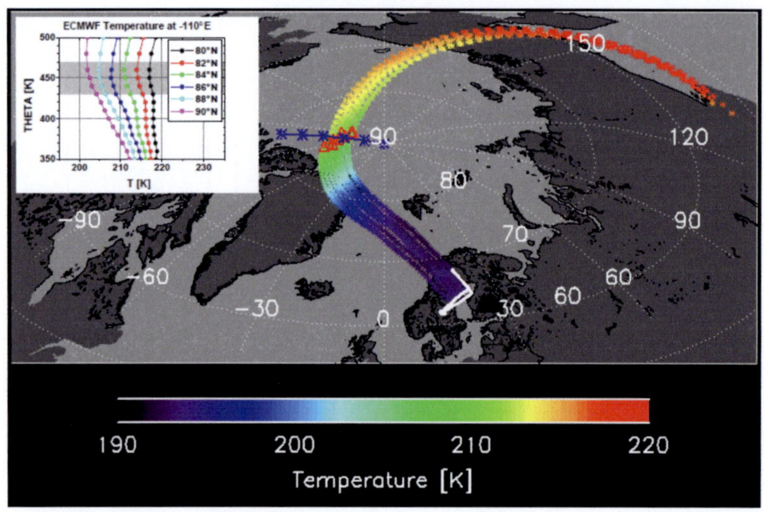

Figure 4.15: Particle sedimentation backward trajectories (coloured asterisks) from CLaMS for 10 particles in the model domain with diameters between 4 to 10 μm on January 25th 2010 at the location of the FSSP-100 measurements. Trajectories are continued by air parcel trajectories prior to the nucleation event. Red triangles marking positions where trajectories reach temperatures of $\sim$210 K. Positions of temperature profiles shown in inset marked by blue asterisks connected by line. Inset: Temperature profiles versus potential temperature. Grey shading indicating altitude range of $\sim$18 to 20 km, referenced to the flight conditions.

the maximum diameters of the particles reported by Fahey et al. [2001] by more than 10 μm. As shown by von Hobe et al. [2012] for a comparable size distribution for the same flight, the number density of the second mode around 14 μm is by a factor $\sim$5 higher than that for the large particle mode reported by Fahey et al. [2001].

The growth conditions of these particles were investigated by analysing particle backward trajectories from the CLaMS model. CLaMS allows the extraction of particle backward trajectories for NAT particles calculated in the simulation at a given time and location. Figure 4.15 shows CLaMS backward trajectories for the largest simulated particles with maximum diameters between 4 and 10 μm found for the FSSP-100 sampling interval on January 25th in the reference run. The simulated particle backward trajectories show the following characteristics:

- The trajectories (coloured asterisks) remain compact during the entire time interval analysed, indicating a homogeneous flow of the overlaying airmasses (particles with different diameter and mass pass certain layers at different time due to different sedimentation speeds).

- All trajectories approach $\sim$T$_{\text{NAT}}$ above the north-east of Greenland two days before arriving at the geolocations of the FSSP-100 measurements. As a consequence, particles with maximum diameters of $\sim$10 μm are obtained in the simulation.

- Going back in time, the tracjectories show further increasing temperatures. About 2.8 days before the FSSP-100 measurements temperatures around 210 K are approached (red triangles).

- Further temperature profiles in the interval from 80°N to 90°N perpendicular to these positions (blue asterisks connected by line) show temperatures well above T$_{\text{NAT}}$ ($\sim$195 K) at the considered altitudes.

The shown backward trajectories indicate that given the stratospheric temperatures (from ECMWF ERA-Interim reananlysis) considered in the simulation the growth of spherical particles with maximum diameters of $\sim$10 μm was possible. This is in contradiction with the FSSP-100 size distribution, yielding particles with maximum sizes of up to $>$30 μm when data processing is carried out assuming spherical particles.

Considering the compactness of the backward trajectories for the particles with diameters between 4 tp 10 μm, strong shear in the flow of the overlaying airmasses allowing larger particles to enter the FSSP-100 measurement interval from very different trajectories appears unlikely. Furthermore, the finding that the temperature profiles perpendicular to the geolocations where the trajectories reach 210 K show all temperatures well above T_{NAT} further reduces the likelihood of such a scenario.

However, the assumption of significantly aspheric particles (i.e. platelets or needles) or particles consisting of loosely packed subunits ('flakes') would allow growth of large particles in a shorter time and might explain the sizes indicated by the FSSP-measurements.

4.3.4 MIPAS-STR measurements and CLaMS simulations of HNO$_3$ redistribution for the flights on January 25th and January 30th 2010

In the following, the HNO$_3$ redistribution patterns observed by MIPAS-STR during the flights on January 25th and January 30th 2010 are compared with the results from the CLaMS simulation termed as 'reference run' (compare Section 4.3.2).

For achieving proper comparisons between the MIPAS-STR retrieval results and the CLaMS simulations, the lower horizontal and vertical resolution of MIPAS-STR is taken into account. The MIPAS-STR retrieval results for HNO$_3$ are typically characterised by a vertical resolution of 1.0 to 1.5 km. Furthermore, airborne IR limb-sounders have typical horizontal resolutions of several tens to a few hundreds of kilometers, as discussed by Ungermann et al. [2012] for the comparable CRISTA-NF instrument. In contrast, the horizontal and vertical resolutions of the CLaMS simulations considered here are $\sim$70 km and $\sim$ 0.5–0.6 km, respectively. Horizontal smoothing of by the MIPAS-STR measurements is considered by the following approach: The CLaMS results are extracted (i) directly at the MIPAS-STR virtual tangent points[6] and also the same positions shifted by the half-distance observer-tangent point (i) towards and (ii) away from the observer and are then weighted by the ratio of 3:1:1. At flight altitude and above, the CLaMS results are extracted directly at the observer

---

[6]I.e. spatial and temporal interpolation of the finer retrieval grid onto the coarser measurement grid.

coordinates and two further locations shifted away from the observer position (by a few ten kilometers, depending on the situation of the uppermost tangent point) into the viewing direction and weighted by the same ratio. This approach allows considering the effects from horizontal atmospheric gradients in the model domain and therefore improves the quality of the comparisons. With respect to the vertical resolution, the extracted CLaMS profiles are smoothed with the averaging kernels of the corresponding retrieved profiles from MIPAS-STR (compare Section 3.2.5). It is pointed out that these approaches for considering for horizontal and vertical smoothing by the MIPAS-STR retrieval do not change the overall outcome of this study. However, the comparability of the simulations and measurements is improved as effects from local atmospheric gradients and structures in the model domain on the comparison are taken into account.

The retrieved vertical cross-sections of temperature and HNO_3 from MIPAS-STR for the flights on January 25th and January 30th 2010 are presented in Figures 4.4 and 4.7 and are discussed in Sections 4.2.1 and 4.2.2. The retrieval results for the flight on January 25th 2010 show temperatures below T_{NAT} around flight altitude and local HNO_3 maxima attributed to renitrification at altitudes around 16 km. Temperatures higher than T_{NAT} and considerable HNO_3 maxima at altitudes around 16 km attributed to renitrification are found for the flight on January 30th 2010. In Figure 4.16, the vertical cross-sections of HNO_3 retrieved from MIPAS-STR (as in Figures 4.4 and 4.7) and extracted from the CLaMS reference simulation are compared for the discussed flights.

For the flight on January 25th, CLaMS reproduces the MIPAS-STR results to a high degree already in the reference scenario: Low HNO_3 mixing ratios are found in sections I, II (between 6:15 and 6:45 UTC) and III around flight alitidue due to significant part of this gas being condensed in PSC particles[7]. Between 6:45 and 7:10 UTC significantly higher HNO_3

---

[7]The individual sections of the flight on January 25th are labeled here different than in Figure 4.4 when comparing with the flight on January 30th for clarity (I-III instead of A-C).

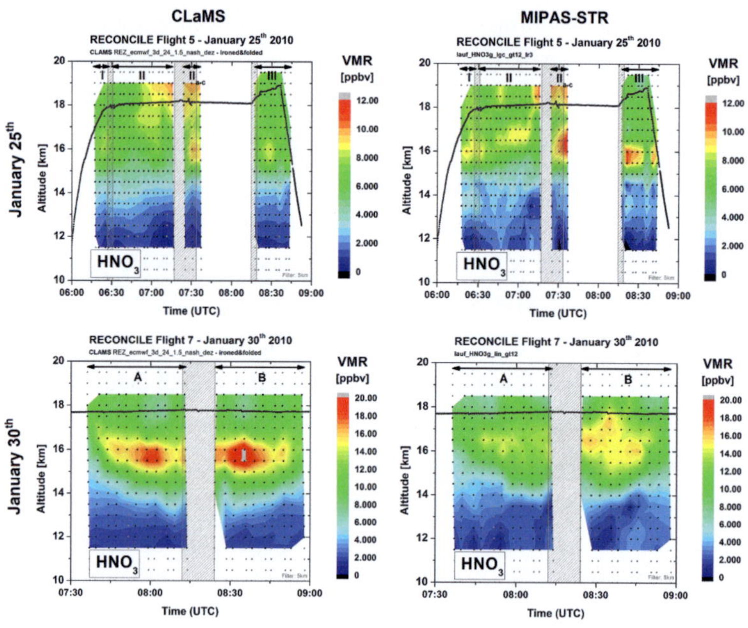

Figure 4.16: Horizontally and vertically smoothed simulated cross-sections of HNO$_3$ extracted from CLaMS corresponding to the MIPAS-STR observations on January 25th and 30th 2010 (left side) and corresponding retrieved vertical cross-sections from MIPAS-STR (right side). Grey lines: Flight altitude of the Geophyisca. Black/grey dots: Retrieval grid. Grid points with nominal vertical resolution of MIPAS-STR retrieval results better than 5 km (black dots) used for interpolation.

mixing ratios are found in the simulation. Around 7:30 UTC, the two overlaying HNO$_3$-maxima found in the MIPAS-STR observations are reproduced in the simulation to a high degree, whereas weak local maxima at altitudes around 16 km found in the MIPAS-STR results in sections I, II and III are hardly indicated by the simulation. However, one to one agreement to such small local structures with such small amplitudes cannot be expected from the comparisons due to constraints from the measurement side (i.e. effects from small scale structures and atmospheric gradients integrated by the measurement), the model domain (limited accuracy of background wind and temperature fields) and the comparison technique.

For the flight on January 30th, CLaMS also reproduces the major structures seen in the MIPAS-STR observations to a high level: Low HNO_3 mixing ratios around flight altitude hint on denitrified air, and both for the north- and south-looking observations (sections A and B) significant HNO_3 maxima through renitrification are found (for this flight, the colour-coding of the MIPAS-STR result is different compared to Figure 4.7 for easier comparisons with the CLaMS result). However, the maximum found in the CLaMS domain for the south-looking observation geometries appears more narrow in vertical direction than in the measurement. Furthermore, considerably higher maximum mixing ratios increased by $\sim$5 ppbv are found compared to the MIPAS-STR result.

From Figure 4.16 it can be seen, that the atmospheric structures seen by MIPAS-STR are reproduced by CLaMS to a high degree already in the reference scenario. However, for the flight on January 30th considerable differences are found between simulation and measurement in the maximum of the renitrification layer. Conclusive comparisons of individual profiles are difficult, as local structures appear different in the model domain for the discussed reasons. However, the MIPAS-STR measurements during the individual flights cover extended parts of the polar vortex, with the measurements covering hundreds of kilometers along flight track and different viewing directions. Therefore, meaningful comparisons are obtained when comparing the whole sets of MIPAS-STR data points from the characteristic flights with the whole sets of the associated CLaMS data points.

4.3.5 Comparison of modelled and measured HNO$_3$ redistribution: The impact of reduced particle settling velocites

In this subsection the comparisons of modelled and measured HNO$_3$ redistribution are discussed for the flights on January 25th and 30th 2010. Different model scenarios with reduced settling velocities of the simulated particles are considered. As pointed out before, the assumption of (i) significantly aspheric particles or (ii) particles with low particle mass density offers an explanation for the large particle sizes observed in-situ during the flight on January 25th 2010.

Table 4.1 summarises the different scenarios with reduced particle settling velocities compared with the MIPAS-STR observations. The reference scenario considers compact spherical particles with the particle mass density described by Hanson and Mauersberger [1988]. The factors for the reductions in settling velocities for aspheric particles in the CLaMS simulations were extracted and converted from the corresponding Figures in Westbrook [2008]. When calculating the relative settling speed reductions, effects from different slip correction factors corresponding to different particle sizes (see Pruppacher and Klett [1997]) on the conversion of the correction factors within a few percent were neglected. For the 'flake-like' particles, the reduction in settling velocities was implemented in CLaMS by using the particle mass densities indicated in Table 4.1, which resulted in the indicated settling speed reductions. For the calculation of the indicated relative settling velocities in Table 4.1 also differences in the slip correction parameters within a few percent are neglected. It is pointed out that this study is not sensitive to estimate the parameters shape and size of the particles involved in the observed denitrification process quantitatively. However, the values given in Table 4.1 representative for the different CLaMS scenarios represent estimates for particle shape and size classes that might explain the particle sizes observed by the in-situ measurements.

Table 4.1: Characteristics of CLaMS scenarios simulating denitrification by NAT particles with reduced settling velocity corresponding to alternative particle shape or morphology. Relative maximum dimension and fall velocity referenced to mass-equivalent particle of the reference scenario ('Compact Spheres').

Name	Mass density (ρ)	Aspect ratio	Relative Maximum Dimension	Simulated Relative Fall Speed
Compact Spheres	1.6 g/cm^3	1 (Sphere)	10 μm	1.00
Platelets	1.6 g/cm^3	1/5	15 μm	0.67
Needles	1.6 g/cm^3	8/1	35 μm	0.67
Flakes I	0.16 g/cm^3	1 (Sphere)	22 μm	0.48
Flakes II	0.004 g/cm^3	1 (Sphere)	74 μm	0.14

Denitrification by potential disk-like particles ('Platelets') was modelled by approximating the fall speed reduction versus a compact spherical particle according to the Roscoe prediction for a horizontally oriented hexagonal disk (extrapolated and converted from Figure 1 in Westbrook [2008]). Thereby, the aspect ratio of 1/5 is the height to diameter ratio. This prediction for horizontally settling particles was chosen considering the work of Katz [1998]. Accordingly, the assumed platelet-like particles with potential maximum diameters of a few tens of microns are expected to show preferentially a horizontal orientation within an angular dispersion of a few degrees. This behaviour is characteristic for disk-shaped particles in this magnitude when settling gravitationally in a flow characterised by low Reynolds numbers (here in the order of 0.01) as the consequence of orienting torques. As the largest expected disk-shaped particles would have diameters of a few tens of microns, such horizontally oriented particles might play a leading role in denitrification. The relative maximum dimension of 15 μm corresponds to the diameter of a hypothetic circular disk-shaped particle with the indicated aspect ratio and an equivalent mass to a compact spherical particle with a diameter of 10 μm in the reference scenario. Also the fall speed is

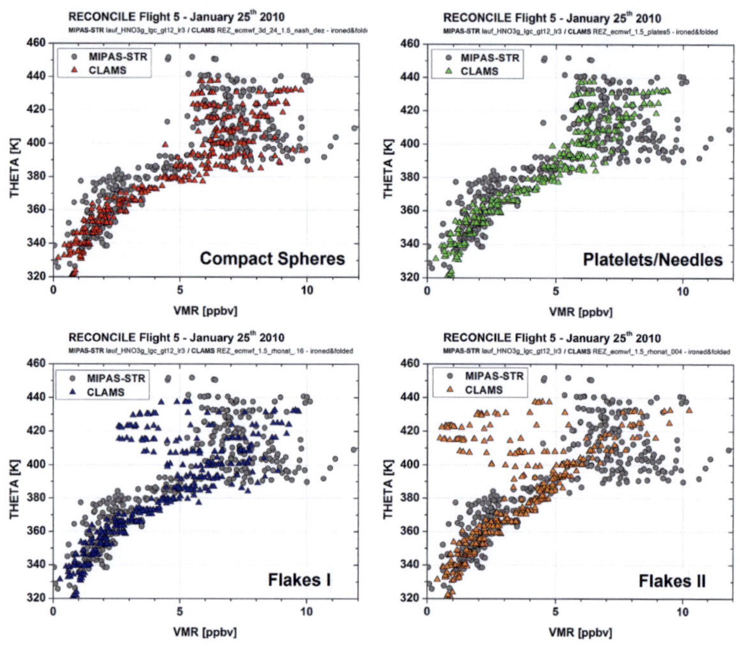

Figure 4.17: Measured vertical distribution of HNO_3 and results from CLaMS simulations considering different particle sedimentation speeds for the flight on January 25th 2010.

referenced to a mass-equivalent particle in the reference scenario. Accordingly, denitrification by disk-shaped compact NAT particles with horizontal orientation was simulated by considering a reduced fall speed of 67 % compared to the reference scenario. The same fall speed reduction corresponds to needle-like particles ('Needles') with an aspect ratio (i.e. height to diameter ratio) of ~8/1 when considering the Hubbard-Douglas prediction for randomly oriented hexagonal needle particles (extrapolated and converted from Figure 3 in Westbrook [2008]). Here, no significant preference orientation is expected for the particles. The indicated maximum dimension of 35 μm corresponds to the height of a hypothetic circular cylinder.

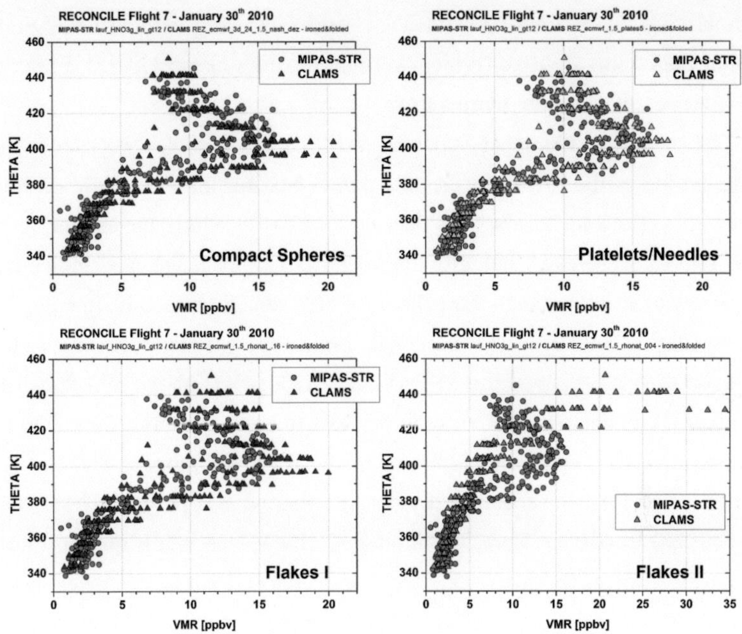

Figure 4.18: Measured vertical redistribution of HNO_3 and results from CLaMS simulations considering different particle sedimentation speeds for the flight on January 30th 2010. Note different scaling of x-axis compared to Figure 4.17.

The scenarios 'Flakes I' and 'Flakes II' consider the fall speed reduction for spherical particles with reduced particle mass densities of 1/10 and 1/400 compared to the value of 1.6 g cm^{-3} considered in the other scenarios. Spherical particles with similar mass compared to the indicated particle for the reference scenario would have diameters of 22 μm and 74 μm, repectively. The respective modelled settling velocities were reduced to about 48 % and 14 % of a mass-equivalent particle of the reference scenario. In Figures 4.17 and 4.18 the results from the different model scenarios are compared with the MIPAS-STR retrieval results. For the plots all data points with the MIPAS-STR retrieval results showing vertical resolutions better than 5 km are used (typical vertical resolutions are 1.0 to 1.5 km for HNO_3) versus potential temperature.

The scenarios 'Compact Spheres' and 'Platelets/Needles' reproduce the distribution of the MIPAS-STR data points best for the PSC flight on January 25th. Both scenarios yield data points mostly within the scattering of the MIPAS-STR result, but with the scenario 'Compact Spheres' indicating a narrow maximum due to renitrification between 380 to 410 K. In contrast, the scenarios 'Flakes I' and 'Flakes II' indicate lower HNO_3 mixing ratios outside the scattering of the MIPAS-STR data points above 390 K, while 'Flakes II' yields lowest mixing ratios close to zero at levels around 420 K.

The MIPAS-STR observations from the flight on January 30th are reproduced well by the scenario 'Compact Spheres' except for the levels between 390 to 410 K. Here, the modelled HNO_3 maximum mixing ratios are overestimated by up to $\sim$5 ppbv. The scenario 'Platelets/Needles' however reproduces the distribution of the MIPAS-STR data points well at all levels. The scenario 'Flakes I' again overestimates the maximum mixing ratios between 380 to 410 K and shows also higher mixing ratios above 420 K compared to MIPAS-STR. The scenario 'Flakes II' shows a distribution rather different from the MIPAS-STR observations without any significant renitrification.

So in summary, the MIPAS-STR measurements on January 25th are reasonably reproduced by the scenarios 'Compact Spheres' and 'Platelets/ Needles'. The reference scenario 'Compact Spheres' slightly better reproduces the scattering of the MIPAS-STR data points in the range 380 to 410 K.

The finding that the scattering within this structure is reproduced to less extent by 'Platelets/Needles' might for example be caused by slightly too slow simulated particle settling speeds, while this parameter allows certain particles to reach layers where instantaneous evaporation happens or not. The scenarios 'Flakes I' and 'Flakes II' both yield distributions different from the MIPAS-STR result as the consequence of strongly reduced particle fall speeds: Particles have more time to equilibrate with the gas phase due to longer retention times at particular levels, resulting in the consumption of a higher fraction of available HNO_3. Due to the rather strong particle fall speed reduction in scenario 'Flakes II', the simulated particles remain almost stationary at given levels and consume most of the available HNO_3.

For the flight on January 30th the MIPAS-STR results are clearly reproduced best by the scenario 'Platelets/Needles', with the simulated data points mostly within the scattering of the observations at all altitudes. The renitrification maximum between 390 to 420 K is overestimated by the scenarios 'Compact Spheres' and 'Flakes I', indicating that this structure is not linearly correlated with the particle settling velocity. This finding points on the importance of the dynamic aspect, for example whether the sedimenting particles are exposed to intermediately present warmer layers as a consequence of certain sedimentation speeds or are advected along different trajectories in horizontal direction. The scenarios 'Flakes I' and 'Flakes II' show significant underestimation of the HNO_3 removal above 420 K due to reduced particle settling velocities, with 'Flakes II' showing hardly any structures from de- and renitrification.

It has to be mentioned that the settling speed reduction in the scenario 'Platelets/ Needles' might also be interpreted as scenario for spherical flake-like particles with only slightly reduced particle mass density. However, such particles would have only slightly larger diameters and therefore cannot explain the size distribution derived from the in-situ measurements during the flight on January 25th 2010. Furthermore it is speculated here, that potential NAT particles in the real stratosphere are predominantly individual crystals due to slow growth in a relatively homogeneous gas phase. Vice versa, the scenarios 'Flakes I' and 'Flakes II' might be interpreted as scenarios for compact platelet- or needle-like particles with rather small or high aspect ratios, respectively. For example, a similar fall speed reduction corresponding to the scenario 'Flakes I' would require disk-shaped platelet-like particles with smallest aspect ratios $\lesssim 1/25$ and are therefore not supported by the results from this study Furthermore it is mentioned, that the shapes of aspheric particles correspond to idealised particle shapes. Such as ice particles (i.e. Libbrecht [2005]), real NAT particles might show different crystal shapes under different growth conditions, and the crystals might have more complex shapes, with consequences for the sedimentation speed. Finally it has to be reminded, that the model simulations discussed here are also based on further assumptions (i.e. nucleation rate of NAT) which also might affect the discussed comparisons.

4.3.6 Discussion: Indications for shape and morphology of large-dimension HNO_3-containing particles

This study involves measurements of MIPAS-STR and simulations from the CLaMS model to study denitrification in Arctic winter 2009/10. Denitrification was modelled by considering reduced settling speeds of potential NAT particles to simulate sedimentation of particles with aspheric shape and non-compact morphology. The results of this study support that the particles found by in-situ measurements during the flight on January 25th indeed contained HNO_3, as large HNO_3-containing particles with sizes in the order of 10 μm characterised by sufficiently fast settling velocities are needed to explain the observed denitrification. Within the applied model scenarios, the assumption of compact aspheric (i.e. platelet- or needle-like) NAT particles with moderate aspect ratios results in considerable agreement between observations and simulations. The assumption of 'flake-like' spherical particles with significantly reduced particle mass densities and also extremely aspheric particles characterised by similar strong reductions in settling velocities are not supported by the results of this study.

While the method of this study is not sensitive to estimate particle shape and morphology quantitatively, the consideration of a fall speed reduction of $\sim$30 % for potential platelet- or needle-like particles versus compact spherical particles yields considerable agreement between the measurements and the model results. The corresponding particles would have maximum dimensions that are more consistent with the particle sizes derived from in-situ particle measurements during the flight on January 25th 2010 than the assumption of compact spherical particles.

Accordingly, the results and indications from this study are summarised as follows:

- Potential NAT particles in the Arctic stratosphere involved in the denitrification process have probably significantly aspheric shapes (i.e. plateles or needles) and therefore reduced settling velocities compared to mass equivalent compact spherical particles.

- The reduction in fall speed significantly affects the vertical patterns of HNO_3 redistribution through denitrification.

- The in-situ particle observations of large-dimension potentially HNO_3-containing particles during the flight on January 25th can hardly be explained by 'NAT rocks' reported by Fahey et al. [2001] due to limited growing time, but by significantly aspheric NAT-particles with large maximum dimensions. The results from this study however are not in contradiction with 'NAT rocks' if growth conditions are appropriate. Considering significantly aspheric shapes, for such particles maximum dimensions exceeding the sizes indicated by the in-situ measurement in Figure 4.14 are expected.

- As in-situ particle measurements during RECONCILE frequently obtained size distributions similar to the discussed measurements from the flight on January 25th 2010 (von Hobe et al. [2012]), such particles might play an important role in the denitrification process.

- The consideration of reduced particle sedimentation speeds corresponding to aspheric shape might improve atmospheric chemistry modelling.

5 Summary and outlook

This work reports on the qualification of the airborne FTIR spectrometer MIPAS-STR and studies on the processes of denitrification and chlorine deactivation, which are important for Arctic stratospheric ozone depletion chemistry and the future evolution of the ozone layer. These studies are based on measurements during the RECONCILE campaign in Arctic winter 2009/10. Essential preconditions for these studies were the optimisation of the MIPAS-STR data processing chain and the characterisation of the calibrated measurements in the current instrument configuration.

Utilising the calibrated measurements of MIPAS-STR, a consistent retrieval strategy allowing quantitative retrievals of atmospheric trace gas mixing ratios and temperature in the UTLS region probed by MIPAS-STR was elaborated. The retrieval allows for resolving mesoscale atmospheric structures and filaments with vertical extensions in the order of one km and horizontal extensions of several tens of kilometers. Such a high vertical and horizontal resolution is consistent to the scales of these structures in the stratosphere and is especially important for comparisons with high resolution atmospheric chemistry models.

It was demonstrated that vertically resolved retrievals are possible even from measurements from inside optically partially transparent PSCs. These retrievals give insight into the chemical trace gas composition inside and below PSCs, which is hardly accessible for comparable satellite- and balloon-borne techniques.

The MIPAS-STR retrieval results were comprehensively validated with available collocated in-situ measurements. The comparisons indicate a high degree of agreement between MIPAS-STR and the in situ instruments, and discrepancies can mostly be explained by the heterogeneous character of the atmosphere sampled remotely by MIPAS-STR in comparison to the very localised in-situ measurements.

Retrieval results were discussed in detail for three flights during the field campaign in the Arctic winter 2009/10 associated to the integrated EU-FP7 project RECONCILE. The vertical cross-sections from MIPAS-STR show chemical and dynamical structures associated with the 2009/10 Arctic polar vortex, including narrow filaments with vertical extensions in the order of one kilometer and patterns from chlorine deactivation and de- and renitrification. The comparison of the correlations between $ClONO_2$ versus CFC-12 from MIPAS-STR and ClO versus CFC-12 from in-situ measurements inside vortex air suggests that between the end of January 2010 and the begin of March 2010 about 90 % of active ozone-destroying chlorine species have been deactivated into $ClONO_2$. This finding supports current understanding of Arctic chlorine chemistry and underlines this important pathway of chlorine deactivation limiting Arctic stratospheric ozone depletion (compare Solomon [1999] and references therein).

The Arctic stratospheric winters show a potential trend towards colder temperatures as a consequence of global climate change (Rex et al. [2006]; Sinnhuber et al. [2011]). Therefore, the fast pathway of chlorine deactivation into $ClONO_2$ might become less effective through more extensive denitrification at colder temperatures, limiting the availability of NO_y in the vertical range sensitive to ozone loss. This process might counteract the decrease of available ozone-destroying chlorine species following from the Montreal protocol and therefore motivate further studies on Arctic winter/spring chlorine deactivation. Quantitative comparisons of the discussed MIPAS-STR measurements and atmospheric chemistry models might follow to study chlorine deactivation under the specific conditions of the polar vortex 2009/10 in detail.

Finally, a combined study on potential denitrification by large particles indicated by in-situ measurements during RECONCILE was carried out based on the MIPAS-STR measurements and simulations of the chemistry transport model CLaMS. Thereby, different scenarios of CLaMS considering denitrification by NAT particles with reduced settling velocities were considered. The reduced settling velocities were considered to simulate denitrification by (i) significantly aspheric particles and (ii) 'flake-like' spherical particles with low particle mass densities, which are both characterised by characteristic reduced sedimentation speeds through increased friction in the stratospheric air. The comparison of the observed and simulated HNO_3 redistribution through denitrification for a PSC flight and a flight under cloud-free conditions at stratospheric altitudes at the end of January 2010 inside the polar vortex suggests that the particles indicated by the in-situ measurements during RECONCILE are reconciled best by assuming moderately aspheric NAT particles. 'Flake-like' NAT particles with low mass densities are not supported by this study. Aspheric NAT particles offer a consistent explanation bringing the in-situ particle measurements, the MIPAS-STR measurements and the CLaMS simulations associated to RECONCILE into a context. The consideration of reduced particle settling velocities of large NAT particles through aspheric shape might help improving atmospheric chemistry modelling.

In summary it could be shown for the first time that detailed quantitative studies on the chemical and dynamical situation of the Arctic UTLS region are possible with MIPAS-STR, even under the harsh measurement conditions inside optically partially transparent PSCs. Retrieval products of MIPAS-STR from the RECONCILE flights will be made available to the scientific community as a part of the project (see https://www.fp7-reconcile.eu). Therefore, this work serves also as a reference for potential further work with the MIPAS-STR data in the RECONCILE framework.

Further results from this work contributed to the PremierEX study (Technical Assistance for the Deployment of Airborne Infra-Red and Millimeter-Wave Limb Sounders during the PremierEX Experiment, Spang et al. [2011]) and the PACD study (PREMIER Analysis of Campaign Data,

Cortesi et al. [2012]), which are both preparatory studies for the candidate Earth Explorer 7 mission PREMIER (PRocess Exploration through Measurements of Infrared and millimeter-wave Emitted Radiation) currently studied by the European Space Agency (ESA) in phase A.

The results of this work provide the basis for potential further studies with MIPAS-STR data:

- Further evaluation of the MIPAS-STR RECONCILE and ESSenCe datasets using the extended data sets of the trace gases discussed here, as well as potential retrievals of further trace gases. Further comparisons with model simulations might be carried out to study chlorine deactivation and its link to denitrification in more detail.

- The MIPAS-STR spectra from PSC-flights during RECONCILE and ESSenCe show characteristic signatures from PSC particles, including a signature similar to the 'NAT-signature' discussed by Höpfner et al. [2006]. This signature might be exploited to obtain information on the shape of potential NAT-particles observed during these campaigns.

- Validation of measurements from satellite-borne limb-sounders (i.e. MIPAS-ENVISAT).

- Evaluation of unprocessed MIPAS-STR data recorded during previous campaigns between 1999 and 2011, including measurements in tropical, mid-latitude and Arctic/Antarctic regions.

- Potential further operation of MIPAS-STR in future campaigns, addressing for example detailed studies of the UTLS region in mid-latitude or tropical regions, model comparisons and validation of other instruments.

Finally, a MIPAS-STR database might be set up to provide the calibrated MIPAS-STR measurements and the retrieval products to the scientific community.

Bibliography

C. E. Blom, M. Höpfner, and C. Weddigen. Correction of phase anomalies of atmospheric emission spectra by the double-differencing method. *Applied Optics*, 35(15):2649–2652, 1996. doi: 10.1364/AO.35.002649.

C. E. Blom, T. Gulde, C. Keim, W. Kimmig, C. Piesch, C. Sartorius, and H. Fischer. MIPAS-STR: Entwicklung eines Instruments für Stratosphärenflugzeuge. In *Statusseminar des Ozonforschungsprogramms*, 1998.

S. Borrmann, B. Luo, and M. Mishchenko. The application of the T-matrix method to the measurement of aspherical particles with forward scattering optical particle counters. *J. Aerosol Sci.*, 31:789–799, 2000.

J. W. Brault. New approach to high-precision Fourier transform spectrometer design. *Applied Optics*, 35(16):2891–2896, 1996.

A. W. Brewer. Evidence for a world circulation provided by the measurements of helium and water vapour distribution in the stratosphere. *Quarterly Journal of the Royal Meteorological Society*, 75:351–363, 1949.

K. S. Carslaw, J. A. Kettleborough, M. J. Northway, S. Davies, R.-S. Gao, D. W. Fahey, D. G. Baumgardner, M. P. Chipperfield, and A. Kleinböhl. A vortex-scale simulation of the growth and sedimentation of large nitric acid hydrate particles. *Journal of Geophysical Research: Atmospheres*, 107(D20):SOL 43–1–SOL 43–16, 2002. doi: 10.1029/2001JD000467.

S. Chapman. XXXV. On ozone and atomic oxygen in the upper atmosphere. *Philosophical Magazine*, 10(64):369–383, 1930. doi: 10.1080/14786443009461588.

U. Cortesi, S. Del Bianco, M. Gai, B. M. Dinelli, E. Castelli, D. Gerber, H. Oelhaf, and W. Woiwode. *PREMIER Analysis of Campaign Data*, volume 4, pages 79–239. Special Issue of IFAC TSSR, 2012. ESA-ESTEC Contract 4000101374/NL/10/CT.

M. de Reus, S. Borrmann, A. Bansemer, A. J. Heymsfield, R. Weigel, C. Schiller, V. Mitev, W. Frey, D. Kunkel, A. Kürten, J. Curtius, N. M. Sitnikov, A. Ulanovsky, and F. Ravegnani. Evidence for ice particles in the tropical stratosphere from in-situ measurements. *Atmospheric Chemistry and Physics*, 9(18):6775–6792, 2009. doi: 10.5194/acp-9-6775-2009.

G. M. B. Dobson, H. H. Kimball, and E. Kidson. Observations of the Amount of Ozone in the Earth's Atmosphere and its Relation to other Geophysical conditions. - part iv. *Proceedings of the Royal Society A*, 129:411–433, 1930. doi: 10.1098/rspa.1930.0165.

A. Dörnbrack, M. C. Pitts, L. R. Poole, Y. J. Orsolini, K. Nishii, and H. Nakamura. The 2009-2010 Arctic stratospheric winter - general evolution, mountain waves and predictability of an operational weather forecast model. *Atmospheric Chemistry and Physics*, 12(8):3659–3675, 2012. doi: 10.5194/acp-12-3659-2012.

G. Dufour, R. Nassar, C. D. Boone, R. Skelton, K. A. Walker, P. F. Bernath, C. P. Rinsland, K. Semeniuk, J. J. Jin, J. C. McConnell, and G. L. Manney. Partitioning between the inorganic chlorine reservoirs HCl and $ClONO_2$ during the Arctic winter 2005 from the ACE-FTS. *Atmospheric Chemistry and Physics*, 6(8):2355–2366, 2006. doi: 10.5194/acp-6-2355-2006.

V. Ebert and J. Wolfrum. *Absorption Spectroscopy*, pages 227–265. Springer, second edition, 2001.

D. W. Fahey, R. S. Gao, K. S. Carslaw, J. Kettleborough, P. J. Popp, M. J. Northway, J. C. Holecek, S. C. Ciciora, R. J. McLaughlin, T. L. Thompson, R. H. Winkler, D. G. Baumgardner, B. Gandrud, P. O. Wennberg, S. Dhaniyala, K. McKinney, T. Peter, R. J. Salawitch, T. P. Bu, J. W. Elkins, C. R. Webster, E. L. Atlas, H. Jost, J. C. Wilson, R. L. Herman, A. Kleinbohl, and M. von Konig. The detection of large HNO_3-containing particles in the winter arctic stratosphere. *Science*, 291:1026–1031, 2001.

J. C. Farman, B. G. Gardiner, and J. D. Shanklin. Large losses of total ozone in Antarctica reveal seasonal ClO_x/NO_x interaction. *Nature*, 315: 207–210, 1985.

H. Fischer and H. Oelhaf. Remote sensing of vertical profiles of atmospheric trace constitutents with MIPAS limb-emission spectrometers. *Applied Optics*, 35(16):2787–2796, 1996. doi: 10.1364/AO.35.002787.

J.-M. Flaud and J. Orphal. *Spectroscopy of the Earth's Atmosphere*. John Wiley & Sons, Ltd, 2011. ISBN 9780470749593. doi: 10.1002/9780470749593.hrs099.

J.-M. Flaud, C. Piccolo, and B. Carli. A spectroscopic database for MIPAS. In *Proceedings of the ENVISAT validation workshop, ESRIN, Italy*, 2002.

J.-M. Flaud, G. Brizzi, M. Carlotti, A. Perrin, and M. Ridolfi. MIPAS database: Validation of HNO_3 line parameters using mipas satellite measurements. *Atmospheric Chemistry and Physics*, 6(12):5037–5048, 2006. doi: 10.5194/acp-6-5037-2006.

M. L. Forman, W. H. Steel, and G. Vanasse. Correction of asymmetric interferograms obtained in Fourier spectroscopy. *Journal of the Optical Society of America*, 56(1):59–61, 1966. doi: 10.1364/JOSA.56.000059.

J. B. Greenblatt, H.-J. Jost, M. Loewenstein, J. R. Podolske, T. P. Bui, D. F. Hurst, J. W. Elkins, R. L. Herman, C. R. Webster, S. M. Schauffler, E. L. Atlas, P. A. Newman, L. R. Lait, M. MÃ¼ller, A. Engel, and U. Schmidt. Defining the polar vortex edge from an N2O:potential temperature correlation. *Journal of Geophysical Research: Atmospheres*, 107(D20):SOL 10–1–SOL 10–9, 2002. doi: 10.1029/2001JD000575.

J.-U. Grooß, G. Günther, R. Müller, P. Konopka, S. Bausch, H. Schlager, C. Voigt, C.M. Volk, and G. C. Toon. Simulation of denitrification and ozone loss for the Arctic winter 2002/2003. *Atmospheric Chemistry and Physics*, 5(6):1437–1448, 2005. doi: 10.5194/acp-5-1437-2005.

J.-U. Grooß, R. Müller, P. Konopka, H.-M. Steinhorst, A. Engel, T. Möbius, and C. M. Volk. The impact of transport across the polar vortex edge on Match ozone loss estimates. *Atmospheric Chemistry and Physics*, 8(3): 565–578, 2008. doi: 10.5194/acp-8-565-2008.

H. Grothe, H. Tizek, D. Waller, and D. J. Stokes. The crystallization kinetics and morphology of nitric acid trihydrate. *Physical Chemistry Chemical Physics*, 8:2232–2239, 2006. doi: 10.1039/B601514J.

D. Hanson and K. Mauersberger. Laboratory studies of the nitric acid trihydrate: Implications for the south polar stratosphere. *Geophysical Research Letters*, 15(8):855–858, 1988. ISSN 1944-8007. doi: 10.1029/GL015i008p00855.

W. N. Hartley. On the probable absorption of solar radiation by atmospheric ozone. *Chem. News*, 42:268, 1880. doi: 10.1098/rspa.1930.0165.

S. Hayashida, T. Sugita, N. Ikeda, Y. Toda, and H. Irie. Temporal evolution of ClONO_2 observed with Improved Limb Atmospheric Spectrometer (ILAS) during Arctic late winter and early spring in 1997. *Journal of Geophysical Research (Atmospheres)*, 112(D11):14311, 2007. doi: 10. 1029/2006JD008108.

J. M. Hollas. *Modern Spectroscopy*. John Wiley & Sons Ltd, The Atrium, Southern Gate, Chichester, West Sussex PO19 8SQ, England, fourth edition, November 2003. ISBN 978-0-470-84416-8.

J. R. Holton. *An Introduction to Dynamic Meteorology*. Academic Press, 4th edition, 2004.

J. R. Holton, P. H. Haynes, M. E. McIntyre, A. R. Douglass, R. B. Rood, and L. Pfister. Stratosphere-Troposphere Exchange. *Reviews of Geophysics*, 33:403–439, 1995.

M. Höpfner, C. E. Blom, G. Echle, N. Glatthor, F. Hase, and G. Stiller. Retrieval simulations for MIPAS-STR measurements,. In W. L. Smith and Y. M. Timofeyev, editors, *IRS 2000: Current Problems in Atmospheric Radiation : Proceedings of the International Radiation Symposium, St. Peterberg, Russia, 24-29 July 2000*, pages 1121–1124, Hampton, Virginia, 2001a. A. Deepak Publishing.

M. Höpfner, C. E. Blom, T. von Clarmann, H. Fischer, N. Glatthor, T. Gulde, F. Hase, C. Keim, W. Kimmig, K. Lessenich, C. Piesch, C. Sartorius, and G. P. Stiller. MIPAS-STR data analysis of APE-GAIA measurements. In W. L. Smith and Y. M. Timofeyev, editors, *IRS 2000: Current Problems in Atmospheric Radiation : Proceedings of the International Radiation Symposium, St. Peterberg, Russia, 24-29 July 2000*, pages 1136–1139, Hampton, Virginia, 2001b. A. Deepak Publishing.

M. Höpfner, B. P. Luo, P. Massoli, F. Cairo, R. Spang, M. Snels, G. Di Donfrancesco, G. Stiller, T. von Clarmann, H. Fischer, and U. Biermann. Spectroscopic evidence for NAT, STS, and ice in MIPAS infrared limb emission measurements of polar stratospheric clouds. *Atmospheric Chemistry and Physics*, 6(5):1201–1219, 2006. doi: 10.5194/acp-6-1201-2006.

C. Kalicinsky. *CRISTA-NF observations in the vicinity of the polar vortex.* PhD thesis, Bergische Universität Wuppertal, Fachbereich C - Mathematik und Naturwissenschaften, 2013.

J. I. Katz. Subsuns and Low Reynolds Number Flow. *Journal of Atmospheric Sciences*, 55:3358–3362, 1998. doi: 10.1175/1520-0469(1998) 055<3358:SALRNF>2.0.CO;2.

C. Keim. *Entwicklung und Verifikation der Sichtlinienstabilisierung für MIPAS auf dem hochfliegenden Forschungsflugzeug M55 Geophysica.* PhD thesis, University of Karlsruhe (TH), Department of Physics, Insitute for Meteorology and Climate Research, 2002. Wissenschaftliche Berichte, FZKA 6729.

L. F. Keyser and M.-T. Leu. Morphology of nitric acid and water ice films. *Microscopy Research and Technique*, 25(5-6):434–438, 1993. doi: 10. 1002/jemt.1070250514.

W. Kimmig. *Das Abtastverfahren der Interferogramme des flugzeuggetragenen Fourierspektrometers MIPAS-STR.* PhD thesis, University of Karlsruhe (TH), Department of Physics, Insitute for Meteorology and Climate Research, 2001. Wissenschaftliche Berichte, FZKA 6665.

A. Kleinert. *Quantifizierung und Optimierung der radiometrischen Genauigkeit des Fourierspektrometers MIPAS-B2.* PhD thesis, University of Karlsruhe (TH), Department of Physics, Insitute for Meteorology and Climate Research, 2003. Wissenschaftliche Berichte, FZKA 6909.

A. Kleinert. Correction of detector nonlinearity for the balloonborne Michelson Interferometer for Passive Atmospheric Sounding. *Applied Optics*, 45:425–431, 2006.

A. Kleinert and O. Trieschmann. Phase determination for a Fourier transform infrared spectrometer in emission mode. *Applied Optics*, 46(12): 2307–2319, 2007. doi: 10.1364/AO.46.002307.

P. Konopka, J.-U. Grooß, G. Günther, D. S. McKenna, R. Müller, J. W. Elkins, D. Fahey, and P. Popp. Weak impact of mixing on chlorine deactivation during SOLVE/THESEO 2000: Lagrangian modeling (CLaMS) versus ER-2 in situ observations. *Journal of Geophysical Research (Atmospheres)*, 108:8324, 2003. doi: 10.1029/2001JD000876.

K. G. Libbrecht. The physics of snow crystals. *Reports on Progress in Physics*, 68:855–895, 2005. doi: 10.1088/0034-4885/68/4/R03.

G. L. Manney, M. L. Santee, M. Rex, N. J. Livesey, M. C. Pitts, P. Veefkind, E. R. Nash, I. Wohltmann, R. Lehmann, L. Froidevaux, L. R. Poole, M. R. Schoeberl, D. P. Haffner, J. Davies, V. Dorokhov, H. Gernandt, B. Johnson, R. Kivi, E. Kyro, N. Larsen, P. F. Levelt, A. Makshtas, C. T. McElroy, H. Nakajima, M. Concepcion Parrondo, D. W. Tarasick, P. von der Gathen, K. A. Walker, and N. S. Zinoviev. Unprecedented Arctic ozone loss in 2011. *Nature*, 478:469–475, 2011. doi: 10.1038/nature10556.

J. J. Marti and K. Mauersberger. Evidence for Nitric Acid Pentahydrate Formed under Stratospheric Conditions. *The Journal of Physical Chemistry*, 98(28):6897–6899, 1994. doi: 10.1021/j100079a001.

G. Maucher. *Das Sternreferenzsystem von MIPAS-B2: Sichtlinienbestimmung für ein ballongetragenes Spektrometer zur Fernerkundung atmosphärischer Spurengase*. PhD thesis, University of Karlsruhe (TH), Department of Physics, Insitute for Meteorology and Climate Research, 1999. Wissenschaftliche Berichte, FZKA 6227.

G. Maucher. personal communication, 2009.

J. C. McConnell and J. J. Jin. Stratospheric Ozone Chemistry. *Atmosphere-Ocean*, 46(1):69–92, 2008. doi: 10.3137/ao.460104.

M. B. McElroy, R. J. Salawitch, S. C. Wofsy, and J. A. Logan. Reductions of Antarctic ozone due to synergistic interactions of chlorine and bromine. *Nature*, 321:759–762, 1986.

MDB. *High-altitude M55 Geophysica aircraft, Investigators Handbook.* Russia, third edition, 2002.

G. Mie. Beiträge zur Optik trüber Medien, speziell kolloidaler Metalllösungen. *Annalen der Physik*, 25(3):377–445, 1908.

L. T. Molina and M. J. Molina. Production of Cl_2O_2 from the self-reaction of the ClO radical. *Journal of Physical Chemistry*, 91(2):433–436, 1987.

M. J. Molina and F. S. Rowland. Stratospheric sink for chlorofluoromethanes - chlorine atomic-catalysed destruction of ozone. *Nature*, 249:810–812, 1974.

S. A. Montzka, S. Reimann, A. Engel, K. Krüger, S. O'Doherty, and W. T. Sturges. *Ozone-depleting substances (ODSs) and related chemicals*, chapter 1. World Meteorological Organization, 2011.

D. P. Moore, A. M. Waterfall, and J. J. Remedios. The potential for radiometric retrievals of halocarbon concentrations from the MIPAS-E instrument. *Advances in Space Research*, 37(12):2238–2246, 2006.

E. R. Nash, P. A. Newman, J. E. Rosenfield, and M. R. Schoeberl. An objective determination of the polar vortex using Ertel's potential vorticity. *Journal of Geophysical Research: Atmospheres*, 101(D5):9471–9478, 1996. doi: 10.1029/96JD00066.

H. Norton and R. Beer. New apodizing functions for Fourier spectrometry. *Journal of the Optical Society of America*, 66(3):259–264, 1976.

H. Norton and R. Beer. Erratum to new apodizing functions for Fourier spectroscopy. *Journal of the Optical Society of America*, 67(3):419, 1977.

T. Peter. Microphysics and heterogeneous chemistry of polar stratospheric clouds. *Annual Review of Physical Chemistry*, 48:785–822, 1997.

T. Peter and J.-U. Grooß. *Polar Stratospheric Clouds and Sulfate Aerosol Particles: Microphysics, Denitrification and Heterogeneous Chemistry.* RSC Publishing, UK, 2012.

D. L. Phillips. A Technique for the Numerical Solution of Certain Integral Equations of the First Kind. *Journal of the ACM*, 9:84–97, 1962.

C. Piesch, T. Gulde, C. Sartorius, F. Friedl-Vallon, M. Seefeldner, M. Wölfel, C. E. Blom, and H. Fischer. Design of a MIPAS instrument for high-altitude aircraft. In *Proc. of the 2nd Internat. Airborne Remote Sensing Conference and Exhibition, ERIM, Ann Arbor, MI*, volume II, pages 199–208, 1996.

M. C. Pitts, L. R. Poole, A. Dörnbrack, and L. W. Thomason. The 2009-2010 Arctic polar stratospheric cloud season: a CALIPSO perspective. *Atmospheric Chemistry and Physics*, 11(5):2161–2177, 2011. doi: 10. 5194/acp-11-2161-2011.

H. R. Pruppacher and J. D. Klett. *Microphysics of clouds and precipitation.* Springer, second edition, 1997.

R. J. Purser and H-L. Huang. Estimating effective data density in a satellite retrieval or an objective analysis. *Journal of Applied Meteorology and Climatology*, 32(6):1092–1107, 1993. doi: 10.1175/1520-0450(1993) 032<1092:EEDDIA>2.0.CO;2.

J. J. Remedios, R. J. Leigh, A. M. Waterfall, D. P. Moore, H. Sembhi, I. Parkes, J. Greenhough, M.P. Chipperfield, and D. Hauglustaine. MIPAS reference atmospheres and comparisons to V4.61/V4.62 MIPAS level 2 geophysical data sets. *Atmospheric Chemistry and Physics Discussions*, 7(4):9973–10017, 2007. doi: 10.5194/acpd-7-9973-2007.

M. Rex, R. J. Salawitch, H. Deckelmann, P. von der Gathen, N. R. P. Harris, M. P. Chipperfield, B. Naujokat, E. Reimer, M. Allaart, S. B. Andersen, R. Bevilacqua, G. O. Braathen, H. Claude, J. Davies, H. De Backer, H. Dier, V. Dorokhov, H. Fast, M. Gerding, S. Godin-Beekmann, K. Hoppel, B. Johnson, E. Kyrö, Z. Litynska, D. Moore, H. Nakane, M. C. Parrondo, A. D. Risley, P. Skrivankova, R. Stübi, P. Viatte, V. Yushkov, and C. Zerefos. Arctic winter 2005: Implications for stratospheric ozone loss and climate change. *Geophysical Research Letters*, 33(23):n/a–n/a, 2006. ISSN 1944-8007. doi: 10.1029/2006GL026731.

O. Riediger, C. M. Volk, M. Strunk, and U. Schmidt. HAGAR – A new in situ tracer instrument for stratospheric balloons and high altitude aircraft. *Eur. Comm. Air Pollut. Res. Report*, 73:727–730, 2000.

C. D. Rodgers. *Inverse Methods for Atmospheric Sounding: Theory and Practice*. Series on Atmospheric, Oceanic and Planetary Physics. World Scientific, 2nd edition, 2000.

L.S. Rothman, I. E. Gordon, A. Barbe, D. C. Benner, P. F. Bernath, M. Birk, V. Boudon, L. R. Brown, A. Campargue, J.-P. Champion, K. Chance, L. H. Coudert, V. Dana, V. M. Devi, S. Fally, J.-M. Flaud, R. R. Gamache, A. Goldman, D. Jacquemart, I. Kleiner, N. Lacome, W. J. Lafferty, J.-Y. Mandin, S. T. Massie, S. N. Mikhailenko, C. E. Miller, N. Moazzen-Ahmadi, O. V. Naumenko, A. V. Nikitin, J. Orphal, V. I. Perevalov, A. Perrin, A. Predoi-Cross, C. P. Rinsland, M. Rotger, M. Šimečková, M. A. H. Smith, K. Sung, S. A. Tashkun, J. Tennyson, R. A. Toth, A. C. Vandaele, and J. Vander Auwera. The HITRAN 2008 molecular spectroscopic database. *Journal of Quantitative Spectroscopy and Radiative Transfer*, 110(9–10):533–572, 2009.

G. N. Shur, V. A. Yushkov, A. V. Drynkov, G. V. Fadeeva, and G. A. Potertikova. Study of Thermodynamics of the Stratosphere at High Latitudes of the Northern Hemisphere on the M-55 Geofizika Flying Laboratory. *Russian Meteorology and Hydrology*, 8:43–53, 2006.

B.-M. Sinnhuber, Stiller G. P., R. Ruhnke, T. von Clarmann, S. Kellmann, and J. Aschmann. Arctic winter 2010/2011 at the brink of an ozone hole. *Geophysical Research Letters*, 2011. doi: doi:10.1029/2011GL049784. in press.

N. M. Sitnikov, V. A. Yushkov, A. A. Afchine, L. I. Korshunov, V. I. Astakhov, A. E. Ulanovskii, Krämer, Mangold M., C. A., Schiller, and F. Ravegnani. The flash instrument for water vapor measurements on board the high-altitude airplane. *Instruments and Experimental Techniques*, 50(1):113–121, 2007. doi: 10.1134/S0020441207010174.

S. Solomon. Stratospheric ozone depletion: A review of concepts and history. *Reviews of Geophysics*, 37(3):275–316, 1999.

R. Spang, J. J. Remedios, and M. P. Barkley. Colour indices for the detection and differentiation of cloud types in infra-red limb emission spectra. *Advances in Space Research*, 33(7):1041–1047, 2004. ISSN 0273-1177. doi: 10.1016/S0273-1177(03)00585-4.

R. Spang, J.-U. Gross, G. Günther., M. von Hobe, C. Kalicinsky, M. Riese, F. Stroh, J. Ungermann, D. Gerber, B. P. Moyna, M. L. Oldfield, J. Reburn, R. Siddans, B. Kerridge, H. Oelhaf, and W. Woiwode. Final Report of the PremierEX Study. Technical report, 2011. ESTEC Contract No. 22670/09/NL/CT.

T. Steck. Methods for Determining Regularization for Atmospheric Retrieval Problems. *Applied Optics*, 41(9):1788–1797, 2002. doi: 10.1364/AO.41.001788.

G. P. Stiller, editor. *The Karlsruhe Optimized and Precise Radiative Transfer Algorithm (KOPRA)*. Forschungszentrum Karlsruhe, 2000. Wissenschaftliche Berichte, FZKA 6487.

G. P. Stiller, T. von Clarmann, B. Funke, N. Glatthor, F. Hase, M. Höpfner, and A. Linden. Sensitivity of trace gas abundances retrievals from infrared limb emission spectra to simplifying approximations in radiative transfer modelling. *Journal of Quantitative Spectroscopy and Radiative Transfer*, 72(3):249–280, 2002. doi: 10.1016/S0022-4073(01)00123-6.

R. S. Stolarski and R. J. Cicerone. Stratospheric chlorine: A possible sink for ozone. *Canadian Journal of Chemistry*, 52:1610–1615, 1974.

O. Sumińska-Ebersoldt, R. Lehmann, T. Wegner, J.-U. Grooß, E. Hösen, R. Weigel, W. Frey, S. Griessbach, V. Mitev, C. Emde, C. M. Volk, S. Borrmann, M. Rex, F. Stroh, and M. von Hobe. ClOOCl photolysis at high solar zenith angles: analysis of the RECONCILE self-match flight. *Atmospheric Chemistry and Physics*, 12(3):1353–1365, 2012. doi: 10.5194/acp-12-1353-2012.

A. Tabazadeh and O. B. Toon. The presence of metastable HNO_3/H_2O solid phases in the stratosphere inferred from ER 2 data. *Journal of Geophysical Research: Atmospheres*, 101(D4):9071–9078, 1996. ISSN 2156-2202. doi: 10.1029/96JD00062.

A. N. Tikhonov. Solution of incorrectly formulated problems and the regularization method. *Doklady Akademii Nauk SSSR*, 151:501–504, 1963. Translated in Soviet Mathematics 4: 1035-1038.

O. Trieschmann. *Phasenkorrektur und Radiometrie gekühlter Fourierspektrometer: Charakterisierung des Instrumentes MIPAS-B2*. PhD thesis, University of Karlsruhe (TH), Department of Physics, Insitute for Meteorology and Climate Research, 2000. Wissenschaftliche Berichte, FZKA 6411.

O. Trieschmann, F. Friedl-Vallon, A. Lengel, H. Oelhaf, G. Wetzel, and Fischer H. An advanced phase correction approach to obtain radiometric calibrated spectra of the optically well balanced balloon borne

Fourier Transform Spectrometer MIPAS-B2. In A. M. Larar, editor, *Optical Spectroscopic Techniques and Instrumentation for Atmospheric and Space Research III*, pages 17–24. SPIE 3756, 1999.

A. E. Ulanovsky, V. A. Yushkov, N. M. Sitnikov, and F. Ravegnani. The FOZAN-II Fast-Response Chemiluminescent Airborne Ozone Analyzer. *Instruments and Experimental Techniques*, 44(2):249–256, 2001.

J. Ungermann, C. Kalicinsky, F. Olschewski, P. Knieling, L. Hoffmann, J. Blank, W. Woiwode, H. Oelhaf, E. Hösen, C. M. Volk, A. Ulanovsky, F. Ravegnani, K. Weigel, F. Stroh, and M. Riese. CRISTA-NF measurements with unprecedented vertical resolution during the RECONCILE aircraft campaign. *Atmospheric Measurement Techniques*, 5(5):1173–1191, 2012. doi: 10.5194/amt-5-1173-2012.

C. Voigt, J. Schreiner, A. Kohlmann, P. Zink, K. Mauersberger, N. Larsen, T. Deshler, C. Kröger, J. Rosen, A. Adriani, F. Cairo, G. Di Donfrancesco, M. Viterbini, J. Ovarlez, H. Ovarlez, C. David, and A. Dörnbrack. Nitric Acid Trihydrate (NAT) in Polar Stratospheric Clouds. *Science*, 290(5497):1756–1758, 2000. doi: 10.1126/science.290.5497.1756.

C. Voigt, H. Schlager, B. P. Luo, A. Dörnbrack, A. Roiger, P. Stock, J. Curtius, H. Vössing, S. Borrmann, S. Davies, P. Konopka, C. Schiller, G. Shur, and T. Peter. Nitric Acid Trihydrate (NAT) formation at low NAT supersaturation in Polar Stratospheric Clouds (PSCs). *Atmospheric Chemistry and Physics*, 5(5):1371–1380, 2005. doi: 10.5194/acp-5-1371-2005.

T. von Clarmann. *Zur Fernerkundung der Erdatmosphäre mittels Infrarotspektroskopie: Rekonstruktionstheorie und Anwendung*. Forschungszentrum Karlsruhe, 2003. Wissenschaftliche Berichte, FZKA 6928.

T. von Clarmann, G. Wetzel, H. Oelhaf, F. Friedl-Vallon, A. Linden, G. Maucher, M. Seefeldner, O. Trieschmann, and F. Lefèvre. ClONO$_2$ vertical profile and estimated mixing ratios of ClO and HOCl in winter Arctic stratosphere from Michelson interferometer for passive atmospheric sounding limb emission spectra. *Journal of Geophysical Research*, 102:16157–16168, 1997. doi: 10.1029/97JD00605.

M. von Hobe, S. Bekki, S. Borrmann, F. Cairo, F. D'Amato, G. Di Donfrancesco, A. Dörnbrack, A. Ebersoldt, M. Ebert, C. Emde, I. Engel, M. Ern, W. Frey, S. Griessbach, J.-U. Grooß, T. Gulde, G. Günther, E. Hösen, L. Hoffmann, V. Homonnai, C. R. Hoyle, I. S. A. Isaksen, D. R. Jackson, I. M. Jánosi, K. Kandler, C. Kalicinsky, A. Keil, S. M. Khaykin, F. Khosrawi, R. Kivi, J. Kuttippurath, J. C. Laube, F. Lefèvre, R. Lehmann, S. Ludmann, B. P. Luo, M. Marchand, J. Meyer, V. Mitev, S. Molleker, R. Müller, H. Oelhaf, F. Olschewski, Y. Orsolini, T. Peter, K. Pfeilsticker, C. Piesch, M. C. Pitts, L. R. Poole, F. D. Pope, F. Ravegnani, M. Rex, M. Riese, T. Röckmann, B. Rognerud, A. Roiger, C. Rolf, M. L. Santee, M. Scheibe, C. Schiller, H. Schlager, M. Siciliani de Cumis, N. Sitnikov, O. A. Søvde, R. Spang, N. Spelten, F. Stordal, O. Sumińska-Ebersoldt, S. Viciani, C. M. Volk, M. vom Scheidt, A. Ulanovski, P. von der Gathen, K. Walker, T. Wegner, R. Weigel, S. Weinbuch, G. Wetzel, F. G. Wienhold, J. Wintel, I. Wohltmann, W. Woiwode, I. A. K. Young, V. Yushkov, B. Zobrist, and F. Stroh. Reconciliation of essential process parameters for an enhanced predictability of arctic stratospheric ozone loss and its climate interactions. *Atmospheric Chemistry and Physics Discussions*, 12(11):30661–30754, 2012. doi: 10.5194/acpd-12-30661-2012.

G. Wagner and M. Birk. New infrared spectroscopic database for chlorine nitrate. *Journal of Quantitative Spectroscopy and Radiative Transfer*, 82 (1-4):443–460, 2003. doi: 10.1016/S0022-4073(03)00169-9.

R. Wagner, O. Möhler, H. Saathoff, O. Stetzer, and U. Schurath. Infrared Spectrum of Nitric Acid Dihydrate: Influence of Particle Shape. *The Journal of Physical Chemistry A*, 109(11):2572–2581, 2005. doi: 10.1021/jp044997u.

A. Werner, C. M. Volk, E. V. Ivanova, T. Wetter, C. Schiller, H. Schlager, and P. Konopka. Quantifying transport into the Arctic lowermost stratosphere. *Atmospheric Chemistry and Physics*, 10(23):11623–11639, 2010. doi: 10.5194/acp-10-11623-2010.

C. D. Westbrook. The fall speeds of sub-100 μm ice crystals. *Quarterly Journal of the Royal Meteorological Society*, 134:1243–1251, 2008. doi: 10.1002/qj.290.

G. Wetzel, H. Oelhaf, R. Ruhnke, F. Friedl-Vallon, A. Kleinert, W. Kouker, G. Maucher, T. Reddmann, M. Seefeldner, M. Stowasser, O. Trieschmann, T. von Clarmann, and H. Fischer. NOy partitioning and budget and its correlation with N2O in the Arctic vortex and in summer midlatitudes in 1997. *Journal of Geophysical Research: Atmospheres*, 107(D16):ACH 3-1–ACH 3-10, 2002. doi: 10.1029/2001JD000916.

G. Wetzel, H. Oelhaf, O. Kirner, R. Ruhnke, F. Friedl-Vallon, A. Kleinert, G. Maucher, H. Fischer, M. Birk, G. Wagner, and A. Engel. First remote sensing measurements of ClOOCl along with ClO and ClONO$_2$ in activated and deactivated Arctic vortex conditions using new ClOOCl IR absorption cross sections. *Atmospheric Chemistry and Physics*, 10(3): 931–945, 2010. doi: 10.5194/acp-10-931-2010.

G. Wetzel, H. Oelhaf, O. Kirner, F. Friedl-Vallon, R. Ruhnke, A. Ebersoldt, A. Kleinert, G. Maucher, H. Nordmeyer, and J. Orphal. Diurnal variations of reactive chlorine and nitrogen oxides observed by MIPAS-B inside the January 2010 Arctic vortex. *Atmospheric Chemistry and Physics*, 12(14):6581–6592, 2012. doi: 10.5194/acp-12-6581-2012.

A. Wiegele, A. Kleinert, H. Oelhaf, R. Ruhnke, G. Wetzel, F. Friedl-Vallon, A. Lengel, G. Maucher, H. Nordmeyer, and H. Fischer. Spatio-temporal variations of NO_y species in the northern latitudes stratosphere measured with the balloon-borne MIPAS instrument. *Atmospheric Chemistry and Physics*, 9(4):1151–1163, 2009. doi: 10.5194/acp-9-1151-2009.

W. Woiwode, H. Oelhaf, T. Gulde, C. Piesch, G. Maucher, A. Ebersoldt, C. Keim, M. Höpfner, S. Khaykin, F. Ravegnani, A. E. Ulanovsky, C. M. Volk, E. Hösen, A. Dörnbrack, J. Ungermann, C. Kalicinsky, and J. Orphal. MIPAS-STR measurements in the Arctic UTLS in winter/spring 2010: instrument characterization, retrieval and validation. *Atmospheric Measurement Techniques*, 5(6):1205–1228, 2012. doi: 10.5194/amt-5-1205-2012.

D. R. Worsnop, L. E. Fox, M. S. Zahniser, and S. C. Wofsy. Vapor pressures of solid hydrates of nitric acid - Implications for polar stratospheric clouds. *Science*, 1993.

Nomenclature

A	averaging kernel matrix
A	raw spectrum
a	averaging kernel trace element
B	rotational constant
BB	blackbody spectrum
C	radiometric gain function
D	force constant
d	factor of degeneracy
DOF	degrees of freedom
E	energy
em	emissivity
f	forward spectrum
g	line shape function
$\hat{H}$	Hamilton operator
h	local vertical grid spacing
I	momentum of inertia

IFG	interferogram
J	rotational quantum number
$\mathbf{K}$	Jacobi matrix
$\mathbf{L}$	regularisation matrix
L	spectral radiance
l	optical path coordinate (atmosphere)
m	mass
P	Planck function
p	pressure
PV	potential vorticity
Q	partition sum
R	transition array element
Re	real part
j	line strength
S	calibrated spectrum
s	optical path coordinate (interferogram)
$\mathbf{S_y}$	variance-covariance matrix of measurement
T	temperature
U	offset spectrum
v	three-dimensional wind-field
W	vibrational quantum number

w	angular velocity
x	parameter (profile) vector
y	measurement vector
η	absorber number density
γ	regularisation parameter
λ	wavelength
μ	reduced mass
ν	frequency
$\tilde{\nu}$	wave number
ϕ	phase function
ψ	wave function
ρ	density
σ	cross section
τ	transmission
Θ	potential temperature
AC	alternating current
AHRS	Attitude and Heading Reference System
CCM	Climate Chemistry Model
CFC	chlorofluorocarbon

CLaMS Chemical Lagrangian Model of the Stratosphere

DC   direct current

ECMWF European Centre for Medium-Range Weather Forecasts

ESSenCe ESa Sounder Campaign 2011

FOV   field-of-view

FTIR   fourier-transform infrared

GPS   Global Positioning System

HCFC   hydrogenated chlorofluorocarbon

ILS   instrumental line shape

IR   infrared

LOS   line-of-sight

NAD   nitric acid dihydrate

NAT   nitric acid trihydrate

NESR   noise equivalent spectral radiance

NILU   Norwegian Institute for Air Research

PSC   polar stratospheric cloud

RECONCILE Reconciliation of Essential Process Parameters for an Enhanced Predictability of Arctic Stratospheric Ozone Loss and its Climate Interactions

MIPAS-STR Michelson Interferometer for Passive Atmospheric Sounding - STRatospheric aircraft

STS   supercooled ternary solutions

UTLS   Upper Troposphere\Lower Stratosphere

UV   ultraviolet

VMR   volume mixing ratio

ZV   zenith view

Acknowledgments

This work was carried out in the framework of the RECONCILE project, which was funded by the European Commission under the grant number RECONCILE-226365-FP7-ENV-2008-1. The RECONCILE project gave me the chance to participate in the RECONCILE campaign in 2010 in Kiruna in northern Sweden, which was a great experience for me. Furthermore, during the project, I had the chance to participate in many exciting international project meetings and conferences, which affected me personally in a positve way and encouraged me in my work. The list of people who encourged me and made my work possible in this manner is long, and probably I will not be able to list all of them. However, this is my attempt:

First of all I want to thank Professor Matthias Olzmann, who kindly agreed to be my supervisor in the Institute of Physical Chemistry. Thank you for good discussions and hosting me in your working group during the last 3,5 years. Also thanks to your kind working group, which made me feel being more than a guest in your group.

In the same manner I want to thank Professor Johannes Orphal, who was my supervisor at the Institute of Meteorology an Climate Research, where I performed the practical work for my thesis. Your excitement for the science we are doing was always encouraging for me and hopefully will be reflected by this work.

My special thanks go to my direct supervisor Hermann Oelhaf, who readily shared his great knowledge and excitement about atmospheric chemistry and the MIPAS-instruments with me. It is great for me to work together with an outstanding person, who made the balloon-version of MIPAS possible and greatly contributed to the related space-project MIPAS-ENVISAT.

Many thanks to Guido Maucher, who introduced me into the world of pointing and helped me to get the IT running, so that I could fully concentrate on my work. Thank you for many helpful discussions and encouraging me in what I was doing.

Furthermore, I want to thank Thomas Gulde, a pioneer of the MIPAS-instruments, who introduced me into the instrument MIPAS-STR and with whom I spent many hours into the late night on the Gephysica, making MIPAS-STR ready for the next flights. Many thanks also to Christoph Piesch, who also contributed to MIPAS-STR from the beginning, and Andreas Ebersoldt, who both also travelled with MIPAS-STR around the world. It was really great to spend exciting campaings with you.

Also I want to thank Cornelis Blom, who brought MIPAS-STR onto the Geophysica and made my work possible. I want to thank you for good discussions and send you my best wishes!

Furthermore, I want to thank Corneli Keim, who introduced me into the MIPAS-STR data-processing. Your work provided the basis for my studies!

Many thanks also to Anne Kleinert, who readily shared with me her knowledge on Level-1 processing. Also many thanks to Michael Hoepfner, who kindly answered my manifold questions on retrievals and error budgets. Thanks also Gerald Wetzel, who helped me with my very early steps in the world of retrievals.

My thanks go furthermore to Felix Friedl-Vallon and all other members of our group for encouraging me in my work. It was great to work together with experts in FTIR spectroscopy and the people who made the MIPAS-instruments possible.

Also I want to thank further people in our institute, who readily shared their knowlege with me. Especially, I want to thank Frank Hase, an expert in spectroscopy and radiative transfer questions, for really helpful discussions. Thanks also to Thomas von Clarmann for a helpful discussion in retrieval questions.

Many thanks go also to the colleagues from IEK-7 at the Forschungszentrum Jülich: Thanks to Marc von Hobe, who coordinated the RECONCILE project and gave 'young' scientists like me the change to participate in scientific conferences, meetings and discussions.

Special thanks also to Jens-Uwe Grooß, who kindly hosted me during my stays in Jülich, when we worked out the comparative studies with CLaMS and MIPAS-STR. Thank you for bringing forward our ideas on dentrification by 'Flakes' and other particles and providing me with the corresponding CLaMS runs.

Many thanks also to the other collegues from Jülich, like Professor Martin Riese, Rolf Müller and Reinhold Spang. I was always feeling like being more than a guest when I came to Jülich.

Furthermore, I want to thank Professor Michael Volk from the University of Wuppertal for helping me understanding our measurements and providing me with the HAGAR data. Thanks also to Fred Stroh from Jülich for providing me with the HALOX-data and for good discussions. Also many thanks to Sergey Khaykin and Alexey Ulanovsky, Central Aerological Observatory, Dolgoprudny, Moscow region, Russia and Fabrizio Ravegnani, CNR-ISAC, Bologna, Italy, for helpful discussions and sharing their in-situ data with me. Thanks also to Hans Schlager from the DLR, Oberpfaffenhofen, for providing the SIOUX-data.

My thanks go also the Stephan Borrmann, Sergey Molleker and the team from the Max-Planck Institute for Chemistry in Mainz for sharing their exciting particle measurements and their knowlege about PSC particles with me.

Thanks also to Kurt Mäki, who kindly hosted us at Kiruna airport and arranged a visit by Julia and me at the Ice-Hotel.

My special thanks go to Lamont Poole, who nicely 'captured the moment' in Budapest when the RECONCILE project ended, by writing the song attached in the following. My best wishes also to your wife and to

Michael Pitts, it was nice to stay with you during the RECONCILE campaign and in Budapest!

Finally, I want to thank my family and my partner Julia Keller for encouraging me in my work during the last years. You, my parents, I want to thank for making my studies in chemistry and my PhD-work possible, as well as for everything you did, making me the person I am glad to be.

Thank you, Susanne, for always encouraging me in what I was doing, and listening to me when I wanted to talk. Thank you for giving me the chance to learn riding and making Dorina a member of our family.

Thank you, Wolfgang, for teaching me in flying and encouraging me in studying chemistry and in many other things. I never imagined, what is possible with a degree in chemistry, as for example working with a stratospheric aircraft. It was definitively right to study chemistry and I am glad that I did so.

Thanks to my sister Elisabeth, who currently is doing a great step, and my brother Ulrich, who currently just like me is keeping a ship on the course. Thanks also to Sierra, who always remembered me that there is more than work.

My very special thanks go finally to my partner Julia Keller. Thank you for your love and for supporting me in everything I did! Together we visited so many nice places, solved so many problems and made so many things reality. And I am sure that there will be many, many more things that we will make reality in future!

The NAT Flake Blues

by Lamont Poole

Nearly thirty years ago, there appeared a hole
In the ozone layer above the South Pole.
It was seen by TOMS and measured at Halley Bay, too.
Some thought it was due to a change in the flow.
Some blamed a gas-phase reaction with NO.
And Solomon and others came up with something totally new.
That chlorine molecules could be set free
By reactions on the surfaces of PSCs.
And when the sun came up, the ozone would be eaten away!

Now, I've got the blues, I've got the NAT flake blues.
Yes, I've got the blues, I've got the NAT flake blues.
Oh, I wish I could lose these old NAT flake blues.

Campaigns were held at both ends of the globe
With remote sensing instruments and in situ probes
To determine if one of these theories might have it right.
From Chile and Antarctica, data were collected.
In Sweden and Norway, results were perfected.
From many long missions deep into the polar night.
A high concentration of chlorine monoxide
Anti-correlated with ozone verified
Freons were the culprits and shouldn't be produced any more.

Now, I've got the blues, I've got the NAT flake blues.
Yes, I've got the blues, I've got the NAT flake blues.
Oh, I wish I could lose these old NAT flake blues.

We've discovered new things here in Project RECONCILE.
Papers will continue to be published for quite a while
On cold aerosols and cloud nucleation on dust.
Refractory aerosols seem to abound
And unexpected large NAT particles were found.
CALIPSO data turned out to be an absolute must.
From Kiruna to Rome to Potsdam to Budapest,
The campaign and science team meetings were the best.
I hope we continue to collaborate as time goes by.

Now, I've got the blues, I've got the NAT flake blues.
Yes, I've got the blues, I've got the NAT flake blues.
Someday we will lose these old NAT flake blues.